PELICAN BOOKS

# EXPLORING THE EARTH AND THE COSMOS

Dr Isaac Asimov was born in Russia in 1920 and emigrated with his family to the United States when he was three. In 1949 he received a Ph.D. from Columbia University and joined the School of Medicine, Boston University, as Professor of Biochemistry. He has published over two hundred books, both fiction and non-fiction, and won a Nebula Award and three Hugo Awards, including one in 1966 for *The Foundation Trilogy* (1963). One of Asimov's main contributions to Science Fiction lore has been his formulation (with John W. Campbell) of the Three Robotic Laws, which he uses as material in his short stories (*I, Robot*, 1956; and *The Rest of the Robots*, 1964) and novels (*The Caves of Steel*, 1954; and *The Naked Sun*, 1957). He has written for children under the pseudonym Paul French. Other works include *Nine Tomorrows* (1959), a history of man's attitudes and discoveries within the cosmos, *The Universe* (Pelican), *Asimov's Guide to Science* in two volumes (Pelican), and two volumes of autobiography, *In Memory Yet Green* (1979) and *In Joy Still Felt* (1980).

# Exploring the Earth and the Cosmos

*The Growth and Future of Human Knowledge*

Isaac Asimov

**PENGUIN BOOKS**

Penguin Books Ltd, Harmondsworth, Middlesex, England
Penguin Books, 40 West 23rd Street, New York, New York 10010, U.S.A.
Penguin Books Australia Ltd, Ringwood, Victoria, Australia
Penguin Books Canada Ltd, 2801 John Street, Markham, Ontario, Canada L3R 1B4
Penguin Books (N.Z.) Ltd, 182–190 Wairau Road, Auckland 10, New Zealand

First published in the U.S.A. by Crown Publishers, Inc.
and simultaneously in Canada by General Publishing Co. Ltd 1982
First published in Great Britain by Allen Lane 1983
Published in Pelican Books 1984

Made and printed in Great Britain by
Richard Clay (The Chaucer Press) Ltd, Bungay, Suffolk
Set in Times

*Dedicated to*
*the memory of*
*Herbert Michelman (1913–1980)*

# Contents

# Introduction

These days we are so used to the broad range of human knowledge in a hundred different directions that we tend to forget how limited we are as individuals.

Forget all that came before you and all that exists beside you and consider what it is that *you* can experience. It isn't much!

To be aware of "all that the eye can see" is to penetrate a distance of, at most, a couple of kilometers. Travel without the help of other people or of inanimate devices and you'll find even a couple of dozen kilometers a daunting task. In fact, until quite recently, most people spent their entire lives within a few kilometers of the place where they were born. Everything else was, to them, hearsay.

What else? How high can you climb? How deep can you plumb? By yourself, a few meters upward into a tree or a few meters downward into a well. You might watch the birds and dream, but dreaming will have to be your master. And the heavenly bodies? They are the very epitome of the unreachable. To "cry for the moon" is the catchphrase for wanting what you cannot have, and the moon is the *nearest* of these bodies.

What can you remember, all by yourself? Your grandparents perhaps, and the stories they told you about how things were when they were young, and about *their* grandparents? A hundred years back, perhaps, and everything dissolves into mist.

What is the largest thing you can experience? A mountain? The smallest thing? A dust mote dancing in a beam of sunlight? The hottest thing? A bonfire? The coldest thing? A bitter winter morning?

In every direction you are limited, and the world closes in on you with its smallness.

But human beings have attributes no other living thing has as far as we know. We have endless curiosity and enormous ingenuity. Myriads of years ago we evolved a complex system of speech by which we could express abstract concepts and transfer them from one to another. Each of us can master not only our own thoughts and insights but those of others. A whole community can pool its gathered ideas and knowledge.

Later, we developed a system of writing, of recording speech, which meant that thoughts and insights could drift outward in space and time, reaching communities far removed from our own and generations yet unborn.

The pool of knowledge and ideas eventually became worldwide and permanent, so that the time arrived when every person had at his disposal all the gains made by those who share the world with us and by those who came before us.

Knowledge grew, first slowly, and then explosively, and the human horizon moved outward in every direction. We now roam the entire world from pole to pole. We climb the highest mountain, plumb the deepest sea, touch the moon, and send our vehicles past Saturn to tell us what they see.

We grasp and study, with equal ease, the largest star and the smallest atom, and consider such mind-bending concepts as black holes and quarks. We probe toward temperatures rivaling the cores of the hottest stars and the depths of the most isolated reaches of unlit space.

In this book I will try to show you the steps by which it all happened, to give you the tale of the heroic voyage of exploration that our species has taken and that we are heir to—and a glimpse also of the distances we have yet to go.

# PART I
## The
## Horizons
## of
## Space

# 1
# The Eastern Hemisphere

## PREHISTORY

On Earth, life began at least 3,500,000,000 years ago, no more than 1,000,000,000 years after the planet had achieved more or less its present shape. Although we may suspect that Earth is not unique as a life-bearing planet, we do not as yet have any evidence that life exists anywhere else, so we must restrict our discussion of the expansiveness of life to our own planet.

Since we are going to deal with human beings primarily and with purposeful expansion of range, there isn't much to say about the expansion of life generally. Life originated in the sea, and for at least seven-eighths of its total duration on Earth, it remained in the sea. The land surfaces of Earth were barren. Why this should be we are not certain.

Perhaps certain key mutations in living cells finally took place that made them capable of protecting themselves from desiccation outside the seas, lakes, and rivers. Perhaps the ozone layer in the upper atmosphere finally formed to the extent that much of the ultraviolet radiation of the sun was blocked. This would mean that land life, exposed to direct solar radiation, would be protected from the dangerous disruption of too energetic wavelengths. Perhaps the moon was captured and consequently the range of tidal action was suddenly increased, and life was washed farther and farther up the shore and the forces of natural selection produced species adapted to periodic desiccations.

Whatever the cause, life invaded the land surface of Earth about

425,000,000 years ago. Following that invasion, life expanded to fill very nearly every conceivable niche on land. Only the polar ice sheets, the highest mountain peaks, and the driest deserts are relatively lifeless.

From beginning to end, life has consisted of tens of millions of different species, sufficiently distinct to be incapable of interbreeding. Of these, most are now extinct, but some 2,000,000 may exist at the moment.

The different species have different ranges. Some are widespread and range over continents and oceans; others are highly restricted and are found only on some particular small island or in some isolated valley. On the whole, the wider the range, the more nearly immune a species is to local change and disaster and the more successful it is by the one criterion that counts from the evolutionary standpoint—survival.

The group of species in which we are most interested, from a self-centered standpoint, are those belonging to the order Primates. The very name of the order is from the Latin word for "first," for no better reason than that the human species is included.

On the whole, the order Primates is not a spectacularly successful one (always excepting human beings). The first primate may have evolved about 75,000,000 years ago, and there are some 200 species now existing, with a range (if we don't count human beings) almost entirely confined to the tropics. What's more, no one species (if, again, we don't count human beings) is to be found over a large fraction of the range. Some groups are confined to sections of South America, some to sections of Africa, some to sections of southeastern Asia, and so on.

Nevertheless, successful or not, the primates did specialize in brainpower. For their size, they had remarkably large and well-developed brains, and as time went on, newly developed primate species produced still larger brains.

The group of species in the order Primates which developed the brain to its highest pitch were those belonging to the family Hominidae. We may refer to individual species of this family as *hominids*, from the Latin word for "man," since all of them, even the most primitive, were more similar to modern human beings than to the modern great apes that are humanity's closest relatives among living species.

The hominids developed an erect posture at least 14,000,000 years ago. This may have taken place in East Africa to begin with, but the hominids in time spread to South Africa and South Asia.

Despite their expanding brain, there could scarcely be any purposefulness in the expansion of the hominid range. As was true of all other species up to that time, the hominids reacted to pushes and pulls--the push of others (of either their own or other species who already filled an area and who directly competed) and the pull of relatively empty areas rich in food just beyond the already-occupied range.

Like other primates, the early hominids did not expand out of the tropics, for they were not adapted to withstand the cold nights and winters of the temperate zones, nor could they find enough food in the winter months.

*Homo erectus* first evolved about 1,500,000 years ago and had a brain midway in size between that of the earliest hominids and that of present-day human beings.

It was *Homo erectus* that made the most significant advance in the history of humanity, for it was this species that first made deliberate use of fire. That was something that no other species had ever done before, and no other species (except for still more advanced hominids) has done since.

The use of fire made it possible for *Homo erectus* to move farther north than earlier hominids had been able to do—certainly to the vicinity of Peking in China, where fossils and also the remains of campfires have been found, and possibly to Europe.

Perhaps as long as 250,000 years ago, large-brained hominids appeared, members of the species *Homo sapiens,* the one to which we belong. They ranged far north into Europe even though the presence of an Ice Age made the climate particularly severe.

Almost from the start, *Homo sapiens,* using fire and animal skins for warmth, and sharpened stone axes and stone-tipped spears as weapons of offense and defense, developed an unprecedented range for a primate.

The early representatives of *Homo sapiens* were not quite like the representatives who live today, but showed differences in skull shape, for instance. We call them "Neanderthal man" *(Homo neanderthalensis).* To the very end of their existence, Neanderthal men were confined to the "World Island" of Europe, Asia, and Africa.

By 35,000 years ago, however, "modern man" had appeared, a variety of *Homo sapiens* to which we belong. This was soon the only kind of hominid to exist. Presumably, modern men were more efficient in war and wiped out the Neanderthals, though there is evidence they may have interbred with them, too.

It was after the appearance of modern man that the hominid range finally extended outside the World Island.

When each of the periodic Ice Ages was at its height, the sea level sank so low, thanks to the water tied up in the gigantic ice sheets that covered Canada, Siberia, and Scandinavia, that Siberia and Alaska were connected by a wide land bridge, parts of which were not covered by ice.

In those days, Siberian hunters pursued mammoths, a now-extinct relative of the elephant, one which was adapted for Arctic life. In the pursuit they crossed over into North America, where, until then, no hominids or, indeed, apes had ever existed. This was some time before 20,000 B.C. By 16,000 B.C., human beings had reached South America, and by 8700 B.C. they were at the Strait of Magellan, and then across it and into Tierra del Fuego, the island to the southeast of the strait. This was the southernmost bit of land to undergo permanent occupation by precivilized human beings anywhere in the world.

Even before this, human beings took advantage of the lowered sea level to cross into New Guinea and Australia.

By then, almost every land area of significant size that wasn't covered by a solid ice sheet carried its population of *Homo sapiens.*

*Homo sapiens* was the first land species to have a worldwide range, and since then it has grown to be the most populous species of its size the world has ever seen. What's more, no one species of life has ever in all the history of the world made up so large a fraction of the total life mass of the planet.

## ANCIENT LAND TRAVEL

And yet although human beings spread over the face of the globe and occupied all the continents of Earth except Antarctica before civilization began, no individual groups of human beings knew any more of Earth's land surface than their own immediate neighborhood, their own town, their own fields.

Even where trade was conducted over long stretches of land, material changed hands many times and no one trader was likely to have traveled far.

It was not till civilization began that the horizon began to spread outward, and then it was for a sad reason, since human beings learned to use military techniques to build empires. Once empires were in existence, some individuals, at least—rulers, soldiers, couriers—gained a general overview of the whole.

The largest of the ancient empires rivaled the modern United States in area. The Persian Empire and the Roman Empire were each close to 7,500,000 square kilometers * in area, though the former had a population of not more than 15,000,000 and the latter not more than 40,000,000. The early empires of India and China were a bit smaller in area but were more populous. The Han Empire of China, which existed in the time of the Roman Empire, had a population in excess of 50,000,000.

The largest of the preindustrial empires was that of the Mongols. Kublai Khan, in 1270, ruled over an area of more than 28,500,-000 square kilometers; his realm was larger in area than that of any contiguous land empire before or since. It stretched from the Danube River to the Pacific Ocean, made up a fifth of all the land area in the world, and had a population of 120,000,000 people.

Through all this, Western Europe had remained comparatively provincial. In ancient times much of it had been part of the Roman Empire, and educated Westerners knew the Mediterranean world at least through their reading and had a notion of the Middle East as far as India.

After the breakup of the Roman Empire in the 400s, however, western Europe grew barbarous, and its awareness was confined within narrow borders by the hostile world of Islam to the east and south.

The first Western European to penetrate Islamic territories in the east and to write an account of his travels was not a Christian but a Jew. He was Benjamin of Tudela, who was born in Navarre on the Franco-Spanish border in the early 1100s. Between 1159 and 1173, he engaged in a journey far to the east, partly out of an interest in trade and partly to visit various communities of Jews. He penetrated to the western border of China itself. His account, however, being that of a Jew, did not much influence the Christian world of the time. Western Europe encountered the conquering Mongols in 1240 but remained outside the Mongol Empire only because at a crucial moment the Mongol khan died and the Mongol armies retired to elect a new one. The existence of the empire, which brought a vast territory under a single strong rule, made overland travel much easier, at least to the extent that there was less danger from anarchic banditry. The West took advantage of this.

Under the sponsorship of the pope, expeditions were sent eastward to convert the Mongols, gain their help against Islam, and obtain assurance of no further invasion of Christian lands. The first

---

* In this book I am using the metric system, standard in virtually all the world except for the United States. A kilometer is equal to 5/8 of a mile; a square kilometer is equal to a little over 3/8 of a square mile. In reverse, a mile equals 1.6 kilometers; a square mile equals 2.6 square kilometers.

of these missions left in the spring of 1245, under the leadership of a Franciscan friar, Giovanni de Piano Carpini, born about 1180. The mission penetrated all the way to the Mongolian capital at Karakorum (in what is now the Mongolian People's Republic) and returned to Europe after a journey that took more than two years.

Nothing was gained from the Mongols, but soon after his return, Carpini wrote an account of his travels, a sober and accurate one. It was the first opportunity Western readers had to read a reasonable account of Central Asia.

He was outdone, however, by William of Rubruck, a French Franciscan friar, born about 1215. King Louis IX of France was in Palestine participating in a crusade and sent a mission under Rubruck to the Mongols on September 16, 1253. Following in Carpini's tracks, eight years after the earlier expedition, Rubruck accomplished as little, but wrote a still better account than that of Carpini.

The climax came with Marco Polo (1254-1324), a Venetian explorer. While he was still a young boy, his father, Niccolo, and his uncle, Maffeo, had gone eastward on a trading mission. They set forth in 1260, at which time the Mongol Empire was at the peak of its power under Kublai Khan.

The Venetians visited Kublai Khan at his summer palace at Shangtu (which became known as Xanadu in Europe). The khan treated them well and sent them back to Europe.

The Polos returned to China in 1275, and this time Marco was with them. Marco caught the fancy of the khan and rose to high position under him, serving as a trusted diplomat. In the khan's old age, however, the Polos felt they could not trust in the favor of his successor, and when they were given the mission of escorting a Mongol princess to Persia, they seized the opportunity of continuing onward toward home. They finally reached Venice again in 1295.

In 1298, Venice was at war with the rival Italian coastal city of Genoa, and Marco Polo, who held a command in the Venetian fleet, was captured. While in a Genoese prison, he dictated the story of his travels. He did not deal so much with personal matters as with a description of the portions of Asia and Africa with which he was reasonably acquainted. The book proved enormously popular, and from that time on, educated Europeans knew of much of the World Island.


ANCIENT SEA TRAVEL

Travel overland, right down to the time of Marco Polo, and for five centuries afterward for that matter, was always difficult. The fastest means of transportation was by horse; there were streams to be forded, forests to be passed through, broken ground to be crossed, mountain ranges to be climbed, supplies to be carried along.

Water, while a barrier to early man, became a highway once ships were invented. Water, after all, was just about level in calm weather, and floating was an easier form of progression than walking, running, or even riding. Early civilizations began along the courses of rivers such as the Nile, the Euphrates, the Indus, and the Hwang Ho, and in each case the river offered a means of travel and of conducting trade. First, there were rafts of lighter-than-water material such as reeds, bark, logs, or inflated skins. Then the rafts were shaped in order to make them more maneuverable. Then they were hollowed into canoes and boats, so that greater loads could be carried within the hollow, since the greater displacement of water by hollowed boats allowed those greater loads to be floated along.

Such rafts and boats could be easily carried downstream by the flowing water, but had to be rowed or poled upstream again. Eventually, sails were used to catch the wind so that the boat could be propelled more quickly in places where the current was slow—or even against the current. By 2600 B.C., the Egyptians had wooden boats, equipped with sails, moving up and down the Nile.

River navigation was relatively easy. One was always in sight of the banks, and in case of trouble one could always sail, row, or even if need be swim to those banks. The situation was different if one ventured out onto the open sea, through the mouth of the Nile, for instance, into the Mediterranean Sea.

There it was possible to be out of sight of land. With nothing but a waste of water out to the horizon in all directions, a mariner's heart would naturally fail him. How would one find land again? In which direction? And what if storms struck and the frail bark was broken and there was no land in any direction to make for?

The need for trade drove people to the sea, however. The Egyptians wanted wood, for instance, which was in short supply in Egypt but plentiful in Lebanon. Logs could be dragged along the coast from Lebanon to Egypt, but that was a prohibitively difficult job. They could be loaded on board ship and sailed or rowed to Egypt, a much easier job. Ships engaged in such trade, however, hugged the coastline as much as they could.

The first people to develop routine sea trade were the early Minoan people of the island of Crete, as early as 2000 B.C. Crete, since it was an island, had to depend on the sea for trade, but there were other islands or continental land in almost all directions, and that helped. The Cretans did learn to risk getting out of sight of land but did not make long journeys.

The first long-range navigators were the Phoenicians, a people living along the coast of what is now Israel and Lebanon. They were the people known in the Bible as Canaanites. "Phoenicia" is a Greek name from the Greek word for "purple," because the chief Phoenician city, Tyre, was known for its production of a purple dye.

The Phoenicians may have been the first to learn to rely on the stars as navigational guides, discovering that the Big Dipper, for instance, lay always to the north, so that one kept it on the right in order to sail west and on the left in order to sail east.

By 1000 B.C., the Phoenicians were venturing out into the western Mediterranean, and the confused tales of what they found there, mellowed by distance and repetition, laid the foundation for the wonder chapters of the wanderings of Odysseus in the *Odyssey*. (Similarly, the exploration of the Black Sea by early Greek traders laid the foundation for the tales of Jason and the Argonauts.)

Not long after 800 B.C., Utica and Carthage were founded by Phoenicians in the land that is now called Tunisia. Carthage, in particular, grew strong and colonized the shores of the westernmost Mediterranean, both north and south.

The Phoenicians were the first Mediterranean people to venture out through the Strait of Gibraltar into the Atlantic Ocean. It was not recognized as what we would today consider the ocean, however. It was thought of as a river flowing around the flat inhabited land area (that portion of the World Island known to the ancient Greeks and Phoenicians).

The Phoenicians explored both to the north and to the south of the Strait of Gibraltar. Somewhere out in the Atlantic, the Phoenicians discovered the "Cassiterides" or "Tin Isles," where they quarried tin ore. This was crucial in the making of bronze, once the sparse tin sources of the eastern Mediterranean had given out.

The Phoenicians kept the location of the Tin Isles a secret in order to maintain their monopoly of the valuable commodity and did it so well that we have no certain knowledge of the location today. It is usually taken to be Cornwall, the peninsula at the southwestern tip of Great Britain, or the Scilly Isles, off the tip of Cornwall, since both possessed tin mines much exploited in later centuries.

The most remarkable tale of Phoenician exploration is that given by Herodotus, the Greek historian. He says that about 600 B.C., an Egyptian king, Necho, sent off a Phoenician expedition to explore the coast of Africa. The expedition, under the Phoenician navigator

Hanno, consisted of sixty ships and 30,000 colonists, and the intention was to found cities. They sailed down the eastern coast of Africa, circumnavigated the continent, and returned to Egypt by way of the Strait of Gibraltar three years later.

Herodotus states emphatically that he does not believe this, because, he says, the Phoenician account includes a statement that when they were sailing past the southern end of the African continent from east to west, the sun lay on the right hand (to the north). Since the sun in the North Temperate Zone, where all the land known to Herodotus was located, lies always to the south, it was that which produced Herodotus' skepticism. And yet the southern tip of Africa is in the South Temperate Zone, and there the sun would indeed always be in the north. The Phoenicians would never have thought of such a seemingly unnatural situation if they had not actually experienced it, which leads one to believe the circumnavigation was a fact.

About 450 B.C., another Phoenician navigator, Himilco, according to a dim account, explored the Atlantic coast northward along what is now Portugal, Spain, and France, and may even have reached Britain. This may have been the usual Phoenician route to the Tin Isles.

As long as the Carthaginian hold on the Strait of Gibraltar was firm, no ships other than Phoenician could venture out into the Atlantic. In the 300s B.C., however, Carthage was engaged in a series of wars with the Greek city of Syracuse in Sicily, and, to some extent, Carthage had the worst of it. The Carthaginian hold on the strait weakened, and about 320 B.C. a Greek navigator, Pytheas of Massilia (the modern Marseilles), slipped out into the Atlantic and sailed northward.

Like Himilco, Pytheas explored the coasts of Portugal, Spain, and France. He definitely reached Britain and seems even to have sailed around it. He refers to the manner in which the Britons threshed their grain indoors (because of the miserable weather in the island). He also described drinks made of fermented grain (beer) and honey (mead). He noted the manner in which the difference in the length of day and night increased as he went farther north, and remarked on the large ebb and flow of the tides (something that did not take place in the nearly tideless Mediterranean).

He either experienced himself or, more likely, was told of places so far north that there were times when the sun did not rise at all and it was dark the entire day. The land he reached that was farthest north he called Thule. It was most likely Scandinavia. From his day, the ocean was recognized more or less for what it was by the western Europeans.

The Phoenician and Greek voyages of exploration did not set a fashion. The tales were dimly remembered and dismissed as fanciful exaggerations or, even if accurate, of no interest to practical men— and the later masters of the Mediterranean were the Romans, who were preeminently practical and who explored only as far as they conquered and no farther.

At about the time the Phoenicians were engaged in their exploratory voyages, still greater feats were performed by the Polynesians on their side of the world. Indeed, considering the technology available to them, the Polynesians were beyond doubt the greatest and most daring navigators the world has ever seen.

The land area in the South Pacific totals nearly 1,000,000 square kilometers. This area is half again as large as Texas, but that land is scattered in 10,000 tiny pieces over an ocean covering nearly a hemisphere. The land area of the South Pacific makes up not much more than 0.5 percent of the ocean in which it is located.

Starting from the East Indies and Australia about 8000 B.C., the Polynesians gradually spread from island to island, covering huge distances in primitive canoelike ships. By A.D. 1000, they had occupied an enormous Pacific triangle of which the corners were New Zealand in the southwest, the Hawaiian Islands in the north, and Easter Island in the east. Their primitive ships had taken the Polynesians crisscross over some 14,000,000 square kilometers of ocean.

Here again, however, the impact on the rest of the world and on later generations was nil. What's more, the Polynesian culture could not maintain close connections among the islands, and isolation came to be the rule.

## MEDIEVAL SEA TRAVEL

In the later centuries of the Polynesian expansion, new navigators arose in Europe in areas that had not been part of the Roman Empire.

One of the areas was Ireland, which the Roman legionaries had never reached. Between 500 and 800, when the Roman Empire in the west had disintegrated and Britain itself had receded nearly to barbarism, Ireland experienced a golden age of culture and learning.

Christianity had come to Ireland in 433 with St. Patrick, and communities of monks preserved learning and even the use of Greek. One of them, St. Brendan, was the first navigator since Pytheas to dare the open Atlantic. He explored the islands off the coast of Scotland and sailed among the Shetland Islands, 200 kilometers north of Scotland.

Pytheas may have reached the Shetlands, but it is possible that St. Brendan went still farther north to the Faeroe Islands, about 320 kilometers north of the Shetlands. Later Irish monks followed in St. Brendan's path, and in about 795, Irish wanderers may have reached Iceland, some 480 kilometers northwest of the Faeroes.

Nevertheless, the Irish explorations of the Atlantic, like those of the Phoenicians and Greeks, had no permanent effect and led to no permanent settlements. The tales were lost in myth and later distortion until it is next to impossible to tell what, in fact, the Irish explorers did and what they are only fabled to have done.

Then came the Vikings of Norway. For a period of about 250 years, beginning about 790, Viking sea raiders terrorized all the coast of western Europe. Vikings occupied most of Ireland and Scotland, putting an end to the golden age of Ireland. They badly ravaged the Anglo-Saxon kingdoms that had been established in Britain after the withdrawal of the Romans. They pillaged the coasts and rivers of what is now France and Germany and even penetrated the Mediterranean, both directly by sea and overland by way of Russia.

They also sailed into hitherto unknown seas. About 870, a Viking named Ottar sailed northward out of sheer curiosity. He said he wanted to see how far north land existed and whether that far-northern land was populated.

Apparently, he succeeded in rounding the northern end of the Scandinavian peninsula, reaching and passing what we now call North Cape, the farthest-northern reach of the European continent. North Cape is at 71.1 degrees north latitude, about 200 kilometers north of the Arctic Circle. Assuming the account to be true, this is the first known crossing of the Arctic Circle by sea.

Ottar sailed beyond North Cape, where the coast trended southeastward along what we now call the Kola Peninsula, and then entered the White Sea.

The Vikings penetrated even farther north than North Cape. In 1194, according to some old records, Vikings came across an island they called Svalbard. This is usually identified with an island group better known to us as Spitzbergen. Even the southernmost point of Spitzbergen is at 76.6 degrees N and is 450 kilometers north of the Arctic Circle. This represents the farthest-northern reach of the Vikings.

The Vikings made even more dramatic journeys westward. Even before Ottar's journey northward, they had reached the Faeroe Islands, where they established the first permanent colony. If any Irish had settled in the islands previously, they had died out, or had left.

The Faeroe Islands were only a stopping place for the Viking seamen. A Norwegian exile, Ingolfur Arnarson, landed in Iceland in 874. Again, any Irish who had ever been there were gone, and a Viking settlement was made and remained there permanently. The Faeroe Islands and Iceland were the first hitherto empty lands to be permanently settled by Europeans in historic times.

From the mountaintops in northwestern Iceland one can dimly make out land on the horizon (about 120 kilometers to the northwest), and in that part of the island there lived Eric Thorvaldsson toward the close of the 900s. He was generally called Eric the Red, from the color of his hair.

In 982, Eric was exiled for three years for some offense and decided to use the time to go exploring. He reached the western land, a forbidding place of ice and desolation, and sailed around its southern tip (some 550 kilometers west of Iceland). The southwestern coast of what turned out to be a huge island seemed somewhat less bleak than the southeastern coast, and Eric judged it capable of supporting a colony.

In 985, Eric was back in Iceland drumming up colonists for his new land, which he called, for promotion purposes, Greenland. This was a misleading name if ever there was one, but the island has kept it to this day. Fourteen ships established a colony in 986, and this colony persisted for over four centuries. At the height of its success, about A.D. 1200, as many as 3,000 Vikings may have dwelt on the island.

Greenland served as a base for explorations still farther west. About 1000, Leif Ericsson, a son of Eric the Red, led a party to a landing somewhere in the North American continent, perhaps on the island of Newfoundland. For some years, Vikings attempted to found a colony in what they called Vinland (the exact location of which is uncertain), but that did not last long.

Unlike Iceland and Greenland, Vinland was not empty of humanity. There were people there whom the Vikings called Skrellings and who were the people who were later called Indians. Their opposition finally wore out the Vinland colonists, who abandoned the area. Nevertheless, the Greenlanders routinely traveled to the North American coast to obtain wood, for trees were absent in Greenland.

Greenland fell on hard times in the 1300s. The Black Death struck Europe in the 1340s and reached Scandinavia and Iceland in 1349. This cut down the exchange of ships that meant life's blood for the

Greenlanders. The last ship sailed from Norway to Greenland in 1367.

Second, Earth underwent a slight cooling trend and Greenland's climate, very poor at best, became so bad as to make agriculture virtually impossible.

The final blow may have been the Eskimos. These were a people of Asian origin who had learned how to make their peace with the Arctic climate and who had developed techniques for living off sea animals.

The Eskimos had established themselves on the shores of Alaska about A.D. 1 and slowly pushed eastward. About 1000, not long after the Vikings reached southwestern Greenland, the Eskimos had reached northwestern Greenland. Steadily, they worked their way southward, and when they finally reached the Viking settlements, competition between the two groups may have been the last straw.

About 1415, the Greenland colony came to an end and the vast island was left to the Eskimos.

Of the Viking explorations, therefore, little came. To be sure, there was Iceland—but no more. The tales of the other lands that those intrepid northern sea rovers had reached did not penetrate the heartland of western Europe, and what whispers of the existence of Greenland and lands still farther west may have reached them were twisted into fables which some believed and some did not.


## ORIENTAL SEA TRAVEL

While the Irish and Vikings dared the cold northern reaches of the Atlantic, and while the Polynesians braved the vast emptiness of the Pacific, the Arabs were sailing the Indian Ocean. The fact that many people in Indonesia and the Philippine Republic are Moslems today bears testimony to the far reach of these traders whose tales, twisted and romanticized, exist still in the stories of Sindbad the Sailor in *The Thousand and One Nights*.

Yet it was China that, in the Middle Ages, was the most technologically advanced nation in the world. For a brief period of time it had been under Mongol domination, but that had not changed the fact. China hovered at the brink of the breakthrough that was to change everything and was to give the world to those who could carry it through.

The magnetic properties of certain minerals had been known since ancient times. The ability of those minerals to transfer their magnetic properties to a steel needle was also known. Somehow (we don't know exactly how or when or by whom) it was discovered that such a magnetized needle, balanced on a pivot and protected from air currents, would take up a north-south position.

We do, however, know *where* it was discovered. It was discovered in China, where the first mention of such things was in the 1000s.

It is clear (at least by hindsight) that such a thing as a pivoting needle (what we now call a "magnetic compass") could be a marvelous aid to navigation. It could point out the north and indirectly, therefore, *all* directions, day or night, in fair or dirty weather. No one would have to depend on sunrise, sunset, the North Star, the Big Dipper, or anything else that bad weather could hide.

This did not fail to occur to navigators. By 1100, Arab traders, whose ships were plying the Indonesian islands, had picked up knowledge of the north-seeking behavior of a magnetized needle and were using a compass to guide them. This much is reported by a Chinese writer of the period.

News of the compass drifted westward, and the first European to mention it that we know of was the English scholar Alexander Neckam. As early as 1180, he spoke of a compass as one of the essentials of the navigator's art.

The world, then, was ready to fall to the first people who were willing to use the compass boldly, to strike out into the far reaches of the sea, and make all the stretch of the ocean their own.

One nation stood on the brink of success, and it was China.

In 1368, the Chinese had driven out the Mongols and were united under the native Ming Dynasty. With a population that had now reached 80,000,000 despite the devastations of the Mongol conquest, China was more populous than all of Europe, was united where Europe was disunited, had a strong and efficient government, was far more cultured and technologically advanced than any other region on Earth, had better ships and more of them, and had been using the compass the longest. What's more, China had an emperor who was fascinated with the seas and an admiral who could handle a fleet better than anyone else in the world at that time.

The emperor in question was Yung-Lo, who became emperor in 1402 and who dreamed of overseas conquests. Admiral of Yung-Lo's war fleets was Cheng-ho, a eunuch who had been born in Yunnan, a southwestern province of China. In 1405, Cheng-ho set sail with 300 ships and 27,000 men.

Cheng-ho's fleet visited Indonesia, Malaya, and Ceylon, establishing Chinese predominance everywhere, defeating native armies and bringing back captured native rulers. In all, Cheng-ho carried through no fewer than seven successful overseas expeditions and sailed as far west as the Red Sea, visiting Mecca and Egypt.

What went wrong?

Emperor Yung-Lo died in 1424, and his successors were not interested in continuing overseas expeditions. China's self-satisfaction (somewhat justified) as the most advanced power in the world caused it to underestimate the importance of intercourse with other powers. It drew back within itself in an imposed self-isolation and left the rule of the sea to others.

China therefore lost the race to become the first world power. Other nations, far smaller, less populous, and less advanced, won out. (To be sure, however, neither the loser nor the winner knew, at this time, that there was a race.)

# 2
# The World as a Whole

## THE COASTS OF AFRICA

Western Europe was the winner, and, to begin with, only a small portion of it, virtually the smallest and least regarded portion.

Western Europe was, at this time, a rather impoverished and backward portion of the civilized world. It became acquainted with the higher standard of living in the East and with many individual Eastern items of culture and refinement, and of comfort too, in the course of the Crusades. Marco Polo's book also impressed them with its tales of the incredible wealth and luxury of "the Indies." Western Europe hungered increasingly for items such as sugar, spices, and silk that could be obtained only from the East.

These items, and others, were, however, staggeringly expensive, since they traveled overland from east to west through many middlemen, each of whom levied a profit. Furthermore, the Ottoman Turks were growing ever stronger along the shores of the eastern Mediterranean. They were hostile to the Christian West and placed barriers in the way of trade.

It occurred to some Westerners that there might well be a sea route to the Indies that would bypass the Turks and all the middlemen and that would allow for direct trade. One person to whom this thought occurred was Prince Henry (1394–1460), the third son of John I of Portugal.

Portugal was the westernmost of the Western European powers, the last in line for the treasures from the East. It faced the Atlantic and was just north of Africa. Having secured its independence from the larger nation of Castile to its east, Portugal looked for adventure and found it in a war in North Africa. In 1415, a Portuguese force took the city of Ceuta on the northernmost tip of what is now Morocco, and Prince Henry distinguished himself in the attack.

Prince Henry was attracted to Africa as a result, and he began to dream of sailing around it in order to reach the Indies. No one knew how far southward Africa extended and whether circumnavigation was possible, but there was the old tale in Herodotus about the Phoenician circumnavigation 2,000 years before. One could but try.

In 1420, Henry founded a center for navigation at Sagres at the extreme southwestern tip of Portugal. This became a haven for experienced navigators, a place where ships were built according to new designs, where new aids to navigation were devised and tested, where crews were hired and trained, and where expeditions were carefully outfitted. Henry never went to sea himself, but his tireless patronage and inspiration have given him the name Henry the Navigator by which he is universally known.

Year after year, Prince Henry sent out ships to sail down the Atlantic coast of Africa, each one trying to go farther than had the one before. By the time of Henry's death in 1460, Portuguese navigators had reached the Gambia River after following the curve of the African coast, generally southwestward, for 3,000 kilometers. It was a terrific achievement but it still represented only one-fifth of the distance to the Indies (though there was, as yet, no way of telling that).

By 1470, the Portuguese seem to have rounded the bulge of Africa and reached the portion of the coast that trended eastward. With great excitement, the navigators leapfrogged each other's efforts, and by 1472 they were 3,000 kilometers farther east than they had been at the mouth of the Gambia River. All seemed to be going well—and then the coast of Africa turned southward again.

The project languished after that monumental disappointment, but in 1481, John II, a grandnephew of Henry the Navigator, came to the Portuguese throne, and he put new life into the project. Within a year of his accession, Portuguese navigators had set sail on explorations that carried them 2,500 kilometers south of the point where the African coast had turned southward again.

John II hedged his bets by also organizing an expedition through the Mediterranean and Red Sea in 1487. It might not be a practical trade route since it would have to pass through Islamic territory, but it might bring back important information.

Under the leadership of Pedro de Covilhão (1450–1545), the expedition moved through Cairo and then traveled to the southern end of the Red Sea at Aden. There Covilhão took a ship that carried him to India. He then sailed back to the eastern coast of Africa, which he explored as far south as the mouth of the Zambesi River in what is now Mozambique.

Covilhão settled in Ethiopia and sent back a full report. By calculation, the continent of Africa couldn't be more than 2,400 kilometers wide at the southernmost points the Portuguese had reached on the western and eastern coasts. It was 6,400 kilometers wide at the northern end. It might be coming to a point, therefore, and the point might not be much farther southward. One more push—

The push was on. In the same year that Covilhão had left, John II sent Bartolomeu Dias (1450–1500) down the western coast of Africa. He traveled farther south than any navigator before him and then was struck by a storm that carried him farther south still. When the storm passed, Dias found himself in the open sea with no sign of land anywhere.

Assuming that the African shore lay somewhere to the east, he sailed eastward and found nothing. He then turned northward and on February 3, 1488, reached the African shore, which, to his surprise, was running east and west. Somewhere the southerly trend must have ended and the coast must have turned eastward, and he had missed the turning point in the storm.

He sailed eastward along the coast, and, after 4,090 kilometers, it began to turn northeastward. Dias was convinced he had passed the southern extremity of the continent and turned back. He found the point where the coastline abruptly left its east-west orientation and became north-south, the point he had missed in the storm. He called it the Cape of Storms, but when he returned to Portugal and made his report, John II renamed it, appropriately, the Cape of Good Hope.

John II, however, who was, more than anyone else (other than Henry the Navigator), the animating heart behind the African project, did not live to see its completion, for he died in 1495. His cousin, Manuel I, succeeded.

Manuel I sent out Vasco da Gama (1469–1524), who left Lisbon on July 8, 1497, rounded the Cape of Good Hope, and kept on going till he reached India on May 19, 1498.

The contact thus made was never lost. Portugal had bypassed the Islamic world. It was a small country, with a population, at that time, of only 1,250,000, but it had two enormous advantages as it faced the rich and populous lands of Asia. In the first place, it had its ships, which could move from point to point at will, in a way that its landbound opponents could do nothing about. By establishing key naval bases, a few ships and men could control vast stretches of coastline.

In the second place, Portugal had cannon on board those ships, and their opponents had nothing to compare with it. Gunpowder had first been discovered by the Chinese, but they had used it chiefly for displays of fireworks and for noisemakers designed to surprise and frighten men and horses in battle. The secret of gunpowder reached Western Europe at a time when the Mongol Empire had made communication east and west easier, and it was not long before the warlike Christians devised cannon with which to slaughter people.

Tiny Portugal, rather than mighty China, thus became the first world power and established the first overseas empire.

## THE NEW WORLD

Portugal did not long remain the only world power, and the route around the southern tip of Africa did not remain the only conceivable sea route to the Indies.

Since the time of the ancient Greeks, European scholars had known that Earth was spherical. In theory, therefore, one could strike out westward from Europe and could sail across the open ocean until one reached the eastern coast of Asia.

To be sure, that meant sailing over vast stretches of open water far from any land, and no one had ever done that. Even the Portuguese in their great exploratory voyages did not like to get too far from the African coast. Still, with the compass well understood, there was no chance of getting lost. One could always turn back.

In this connection there arose the question of how large Earth might be. The ancient Greek geographer Eratosthenes (276–196 B.C.) had, about 240 B.C., estimated that Earth was (in modern measure) 40,000 kilometers in circumference. He was right, and this meant that the distance from the western coast of Europe westward to the eastern coast of Asia was about 25,000 kilometers. No ship built in the 1400s could possibly carry enough provisions to cross that width of open water.

Later Greek geographers, however, incorrectly settled on a lower figure for the circumference of Earth, and many geographers of the 1400s felt Earth's circumference to be no more than 29,000 kilometers altogether. What's more, Marco Polo had incorrectly placed the eastern shore of Asia farther east than it really was. If the mistake of

Earth's too-small size was added to the mistake of Asia's too-far-east coastline, it turned out that the westward voyage from Europe might not be more than 4,800 kilometers (less than a fifth of what it really was).

Convinced that this small distance was correct, Christopher Columbus, who had been born in Genoa, Italy, in 1451, and who had gone to sea as a teenager, dreamed of reaching Asia by sailing westward. He tried to sell the notion to John II of Portugal, but John II wasn't buying. He believed the westward journey would be far greater than 4,800 kilometers, and he preferred to stick with the African project.

Meanwhile, the nation of Castile, on Portugal's east, had vanished from the map. Its queen, Isabella, had married Ferdinand II of Aragon, the nation still farther east, and the two monarchs ruled over the combined nation, now called Spain. What's more, in 1492, Ferdinand and Isabella conquered Granada, the last stronghold of the Moslems who, seven and a half centuries before, had conquered all of Spain.

Anxious to demonstrate the new power of Spain, and somewhat envious of Portuguese achievements at sea, Ferdinand and Isabella decided to back Columbus cautiously—at least to the extent of giving him three small and fairly worthless ships and offering him a crew of convicts who were willing to volunteer for the dangerous voyage as the price of freedom. The total cost of the expedition, which would prove to be the most momentous in history, has been estimated at anywhere from $16,000 to $75,000, not very much even for those days.

On August 3, 1492, Columbus, with a total crew of ninety on all three ships, left Palos, a port in southern Spain just 50 kilometers east of the Portuguese frontier.

Columbus sailed for the Canary Islands, which were in Spanish hands, and then, on September 6, 1492, took off into the unknown. He did well to travel southward first, since this enabled him to take advantage of the trade winds, which blew him westward on his way. Had he tried to make the crossing farther north he would have had to face the westerlies, which would be blowing in the wrong direction.

For several weeks, Columbus's ships sailed steadily westward. Amazingly it was an entirely smooth passage, the smoothest on record. Not once in all those weeks was there a storm—which was fortunate indeed, for Columbus's three hulks would very likely have foundered in a real storm.

In all that time there was nothing but wind and waves until, on October 12, 1492, the cry of "Land" was raised. The land sighted was a small inhabited island, and Columbus, who was convinced he had reached the Indies, called its inhabitants Indians, a misnomer that remains to this day. Columbus named the island San Salvador ("Holy Savior"), but the name dropped out of use, and, astonishingly enough, the identity of the island is no longer certain. It is usually identified with one of the islands of the Bahamas, the one called Watling's Island after an English pirate, John Watling.

Again, because of Columbus's certainty that the islands he first discovered were part of the Indies, islands off the American coast are called the West Indies to this day.

Columbus was back in Palos on March 13, 1493, and was suddenly the most famous man in Europe. Navigators of all kinds, sponsored by different nations, flocked across the Atlantic, but Spain was first in the field and had become the second world power.

One of the navigators was an Italian named Amerigo Vespucci (1451–1512). The Latinized form of his name was Americus Vespucius. From 1497 onward, he led expeditions that explored the coast of a continental area to the south of the islands discovered by Columbus.

To Vespucci it did not seem that the new lands could possibly be the Asian coast that Marco Polo had described. Columbus thought it was and maintained that point of view till his death in 1504, but Vespucci argued that Eratosthenes was right, that it was 25,000 kilometers from Europe westward to Asia, that the land that had been discovered was a "New World" beyond which lay another ocean. It was only the unforeseen existence of these hitherto unknown continents that had kept Columbus from perishing in a futile effort to reach distant Asia.

Vespucci's views were accepted by a German geographer, Martin Waldseemüller (1470?–1522). In 1507, Waldseemüller published a map showing the new continent as existing by itself and not as part of either Europe, Africa, or Asia. He proposed that it be named America in honor of Vespucci, who had first recognized it as a new continent.

The name was instantly popular and was soon in universal use. At first it was applied exclusively to the southern portion of the New World, for the northern portion might still have been attached to Asia. Eventually, though, the northern portion was also recognized as separate. It became North America, while the southern portion became South America.

## AROUND THE WORLD

Postulating a second ocean is one thing, seeing it is another. The first sight came to a Spanish explorer, Vasco Nuñez de Balboa (1475–1517). He had settled in what is now Panama and, in 1510, heard rumors of tribes with much gold living in the interior. He decided to look for that gold, and on September 1, 1513, headed a party of men who moved into the interior. None of the Europeans knew at the time how narrow Panama was. On September 25 they climbed a last hill and found themselves staring at the limitless expanse of what seemed to be an ocean. Balboa called it the South Sea because at that point it lay to the south of the shoreline.

Balboa's discovery strengthened Vespucci's suggestion that the true Asia lay far to the west.

This was a point made by a Portuguese navigator, Ferdinand Magellan (1480–1521). He had been wounded in a Portuguese war against the Moroccans in North Africa, but had been unjustly accused of trading with the Moroccans and had been denied a pension. Hot with feelings of injustice, he then offered his services to Spain.

The pope had assigned the lands gained by eastward voyages to the Portuguese and the lands gained by westward voyages to the Spaniards. Magellan pointed out that if Spain continued westward past the Americas, it would reach Asia without having journeyed east and could trade with the Indies without violating the letter of the pope's injunction.

The new Spanish king was Charles I, the son of Ferdinand and Isabella (and soon to be better known as the Emperor Charles V of the Holy Roman Empire). Charles V agreed to sponsor the voyage, and Magellan left Spain with five ships in August 1519.

Magellan sailed to the eastern bulge of South America (the modern Brazil), which was then in Portuguese possession, and then sailed southward down the continent looking for some sea passage that would carry his ships through the continent into the second ocean. The expedition had to travel dishearteningly far southward until, on October 21, 1520, it located the passage, still called the Strait of Magellan to this day.

For thirty-eight days the five ships made their way through abominable weather along the 525-kilometer length of the strait. All five ships survived, and on November 28 they sailed out into a smooth and sunny ocean which Magellan, in gratitude, named the Pacific Ocean.

The worst lay ahead, however. The Pacific was as large as might be expected from Eratosthenes' determination of Earth's circumference, but the difficulty was that it was also remarkably free of land. For ninety-nine days, the little fleet traveled onward without any sign of land. Provisions gave out, and the crew had to gnaw leather and mix sawdust with their rations to have something to chew on, until on March 6, 1521, nearly dead, they reached the island of Guam. There they caught their breath and sailed westward to what afterward came to be known as the Philippine Islands. There, on April 27, 1521, Magellan died in a skirmish with the natives.

What was left of the expedition managed to get away from the Philippines in two remaining ships and made it to the Indonesian islands, which the Portuguese had already reached.

Of the two ships, one tried to return across the Pacific and never made it. The final ship, the *Victory,* under the captaincy of Juan Sebastian del Cano (1460–1526), continued westward, rounded the Cape of Good Hope, and returned to Spain with eighteen men aboard on September 9, 1522.

If the loss of life can be set aside, the one surviving ship carried enough spices to make the whole voyage a highly profitable undertaking.

The *Victory* carried the first human beings to circumnavigate Earth, and this was a crucial point. Not only did it prove finally that Eratosthenes' estimate of Earth's size was correct, but also it showed that Earth possessed a global ocean in which the continents were set as huge islands.

More than that, with this circumnavigation of the world, humanity's horizon expanded, for the first time, to the point where it included all of Earth. For the first time, people could think of Earth globally and could know the relative positions of the continents upon it.

And because it was the Europeans who had the ships and the cannon and the navigational expertise, it was the Europeans who controlled the world for four and a half centuries after the voyages of Columbus and da Gama.

## THE PACIFIC OCEAN

Knowing the globe as a whole is far from knowing every detail of the globe. The navigators who plied the oceans with increasing self-confidence in the 1500s and 1600s constantly expected to find new lands as Columbus had done.

There was not much chance that spectacular land discoveries

could be made in the relatively narrow confines of the Atlantic or Indian oceans, but the Pacific Ocean offered more hope. Between the Americas and Asia lay an ocean which, if it was empty, was 188,000,000 square kilometers in area, one-third the entire area of the globe.

If the Pacific Ocean was subtracted, the rest of the globe was land and water in the proportion of 3 to 5. If this was true of the Pacific area also, there should be 70,000,000 square kilometers of land there, an area which, if concentrated in a single continent, would be nearly as large as Asia and Africa combined. And even if that was an overestimate there should still be something respectable there.

In 1606, a Spanish navigator, Luis Vaez de Torres, circumnavigated the large island of New Guinea and showed that it, at least, was not part of a South Pacific continent. The waters to the south of the island are still called Torres Strait in his honor.

Torres had missed seeing land to the south of Torres Strait, but there were reports that such land existed.

In the 1600s, Portugal was in decline and the enterprising Dutch were taking over the Indonesian islands. The Dutch governor-general of the islands, Anton Van Diemen (1593-1645), sent out an exploring expedition under Abel Janszoon Tasman (1603-59) to check on possible land south of New Guinea.

Tasman left Batavia (now Djakarta) on August 14, 1642, with two ships and in the course of ten months managed to circumnavigate a large island the size of the United States without ever seeing it. He did discover a smaller island to its southeast which he called Van Diemen's Land, but which was eventually named Tasmania in his honor. He also discovered the southern island of New Zealand, and the Fiji Islands. On a second expedition in 1644, he did spy the coastline of the large island he had earlier missed, and he named it New Holland. He was not able to show, however, whether it was a single island or several, nor how large it was.

Other navigators, French and English, sailed the Pacific without finding any continents, and the job was finally completed by the English navigator James Cook, perhaps the greatest sea explorer in history.

Cook had the advantage of technology. Until Cook's time, longitude could not be determined, because an accurate timepiece was lacking. Good clocks of any kind did not exist until the Dutch physicist Christiaan Huygens (1629-95) invented the pendulum clock in 1656, but no pendulum clock could keep time on the swaying deck of a ship. In 1713, the British Parliament offered a series of prizes up to 20,000 pounds for a timepiece that would keep time accurately on board ship. The prize was won by an English instrument maker, John Harrison (1693–1776), who constructed his fifth and best watch in 1765.

Then, too, a Scottish physician, James Lind (1716–94), had found in 1747 that citrus fruits prevented scurvy, a disease which routinely disabled men on long ocean voyages, since they were then confined to a monotonous diet of hardtack and salt pork. (The disease resulted from a lack of vitamin C and had first struck ships' crews during Vasco da Gama's voyage.)

Cook, with chronometers and citrus fruits, could scour the Pacific efficiently, and did so in three voyages. In his first voyage, from 1768 to 1771, he carefully charted all the coasts of the two islands of New Zealand and of the entire eastern coast of New Holland, which came to be called Australia and which Cook recognized as a single island large enough to be called a continent.

In his second voyage, from 1772 to 1775, he sailed back and forth over the South Pacific and circumnavigated the world in far-southern latitudes. He found no major landmasses.

In his third voyage, from 1776 to 1779, he sailed into the North Pacific and showed there were no major landmasses there, either. He did not survive the third voyage. On February 14, 1779, he was killed in a brief battle with natives on the Hawaiian Islands, which he was the first European to reach.

Cook's voyages showed that with the possible exception of the polar areas, which were in any case uninhabited and, in the ordinary sense, uninhabitable, no undiscovered landmasses of any size existed. Australia was the last habitable continent to be discovered, and thereafter small islands of one sort or another might be stumbled upon, but nothing more.

It was clearly demonstrated that the Pacific was indeed an enormous ocean, empty of significant land, taking up an area one-third of Earth's surface. Thanks to it, the ocean covers 70 percent of Earth's surface and the land area only the remaining 30 percent.


## THE ARCTIC COAST

If the polar regions were uninhabited and uninhabitable, was there any point in exploring them and in expanding the human horizon in that direction? Practical men might think not, but circumstances drove them in that direction.

After Magellan's voyage there were two sea routes from Europe to the Indies. There was the Southeast Passage, pioneered by da Gama and under Portuguese control, and there was the Southwest Passage, pioneered by Magellan and under Spanish control.

Other European nations such as England, France, and the Netherlands wished to trade and were anxious to seek out other sea routes. There were two possibilities: One would require the skirting of the northern coasts of Europe and Asia (the Northeast Passage) and the other the skirting of the northern coast of North America (the Northwest Passage).

In 1553, the English navigator Sir Hugh Willoughby set out with three ships to reconnoiter the Northeast Passage. The ships rounded the northern edge of Scandinavia as Ottar had done seven centuries before.

Willoughby crossed the Arctic Circle, and his ships then sailed onward about 1,000 kilometers to the east before deciding that it was too late into the winter season to continue. They turned back. Two of the ships, with Willoughby in command, found a harbor on the Kola Peninsula along the Arctic shore just east of Scandinavia. They were forced to spend the winter there and died of the cold.

The third ship, under Richard Chancellor, had been separated from the other two in a storm, fortunately for itself, and had made its way past the edge of the Kola Peninsula into the White Sea. There it found the Russian seaport of Arkhangelsk. Russians took the Englishmen overland to Moscow, where they were greeted by the Russian czar, who at that time was Ivan IV (the Terrible).

The Dutch also made an attempt. In 1594, a Dutch navigator, Willem Barents, left Amsterdam with two ships and explored the stretch of ocean to the north of Scandinavia and western Russia, a stretch still called the Barents Sea in his honor. He discovered several islands of considerable size, including Novaya Zemlya and Spitzbergen, which Ottar may have spied seven centuries before but which had been forgotten.

In the winter of 1596–97, Barents, like Willoughby before him, was frozen in place; in his case, in Novaya Zemlya. There were sixteen crewmen and a cabin boy. The cabin boy died and Barents did not long survive the coming of spring, but the other fifteen made it back to safety, the first Europeans to survive an Arctic winter so far north.

The misadventures of Willoughby and Barents caused Western Europeans to suspect that the Northeast Passage was not really practical, and efforts in that direction ceased.

For the Russians in Eastern Europe, however, it was another matter. They could venture eastward by land, and even if the going was hard there were rewards en route in the form of furs. It had been the fur trade that had lured the Russians as far north as Arkhangelsk in the first place.

In 1581, the Stroganovs, a Russian family that had made a fortune in the fur trade, employed a Cossack named Yermak Timofievich to explore eastward. Yermak conquered a Mongol kingdom east of the Urals. It was named Sibir, and the name (Siberia in English) came to be applied to the entire northern third of Asia.

Other fur traders followed, and by 1648, a Cossack named Semyon Ivanov Dezhnev had reached the far-eastern end of the Asian continent. In less than seventy years, Russia had stretched its dominions eastward thousands of kilometers across the full width of Asia and gained a land it still retains today.

This Russian feat showed that the northern shores of Russia and Siberia never dipped southward but remained north of the Arctic Circle for many thousands of kilometers. If there was any doubt that the Northeast Passage was not practical, that doubt was removed.

But what about the Northwest Passage? Most of the explorations that had followed Columbus's first voyage were toward the south—to Florida and beyond. There were, however, exceptions.

The Italian navigator John Cabot (1450–98), who sailed in the pay of England, touched Newfoundland in 1497. The Portuguese navigator Gaspar Corte Real (1450–1501) spied the coast north of Newfoundland in 1501, and gave it the Portuguese name Labrador that it still carries. (The word means "slaves," and it came about because Corte Real picked up some of its inhabitants and carried them off as slaves.)

These sightings, however, filled in very little of the map. Elsewhere almost anything might exist, almost any hope might be fulfilled.

It was the French who made the first systematic attempt at finding a Northwest Passage. In 1524, they outfitted an expedition under the Italian navigator Giovanni da Verrazano (1485–1528). He explored the eastern coast of North America from the Spanish-controlled lands in the south to the northern peninsula now known as Nova Scotia. He investigated all the major inlets in case they should turn out to be straits leading through to the Pacific. He was the first European to enter what is now New York Bay, for instance.

Verrazano's work showed clearly that it was hopeless to look for the Northwest Passage south of Nova Scotia.

A French navigator, Jacques Cartier (1491–1557), came west in 1534 and found an opening well to the north of Nova Scotia. It lay between Newfoundland and Labrador, and is now called the Strait of Belle Isle. He went through it into what seemed a wide inlet of the ocean. Since he entered it on August 10, a day dedicated to St. Lawrence, it is now called the Gulf of St. Lawrence. It turned out to be the outlet of the St. Lawrence River and did not lead to the Pacific.

If a Northwest Passage existed at all, Cartier's expedition demonstrated that it would have to be farther north than Labrador. Considering Labrador's frigid climate, this meant that the Northwest Passage, like the Northeast Passage, would involve Arctic waters. Yet perhaps there might be just a short stretch through the Arctic and then a curve southward into milder seas. This possibility made it worth looking for.

In 1576, an English navigator, Martin Frobisher (1535–94), sailed to North America with three ships and thirty-five men. One of the ships, with eighteen men aboard, reached Labrador. Frobisher himself went northward with the other two ships and touched the large island now known as Baffin Island, but located nothing that seemed to be promising as a beginning of a Northwest Passage. On a later voyage, in 1578, Frobisher caught a glimpse of the southern tip of Greenland, where, a century and a half before, the Viking colony had come to an end. With this new discovery, Greenland entered and remained in the consciousness of European scholarship and navigation.

The next major attempt was made by an English navigator, Henry Hudson, who, in 1610, with one ship and a crew of nineteen men, found a waterway (now called Hudson Strait) between Baffin Island and Labrador. He passed through and entered a large body of water that extended well southward, but it was not the Northwest Passage at all. It was a landlocked body of water now called Hudson Bay.

Hudson's ship was frozen in for six months during the winter of 1610–11 in the southernmost section of the bay, a region now called James Bay (after the English king James I). When the ice broke, he wanted to continue exploring, but his crew had had enough. They set him, his son, and seven loyal crew members adrift and returned to England. Hudson and the others froze to death.

In 1615, the English navigator William Baffin (1584–1622) explored the waterway between Baffin Island (named for him) and Greenland, a waterway now called Baffin Bay. He followed it to the islands that lay to the north of Baffin Island and nosed into the narrow straits that separated them. He did not consider them practical waterways and returned.

Thereafter it was decided that the Northwest Passage was as impractical as the Northeast Passage.

But by that time it really didn't matter. Spain and Portugal had declined to the point where they could no longer monopolize the sea lanes. Any nation could now sail the ocean freely, and in actual fact, the English and the Dutch between them now carried on most of the sea trade of the world.

# 3
# The Interiors and the Poles

## INSIDE NORTH AMERICA

With the investigation into the Northeast and Northwest passages, and with the exploration of the Pacific Ocean by Captain Cook, the major sea voyages of exploration were over. The ocean and the continental coastlines were known. What about the continental interiors, however?

It was, after all, harder to travel overland than oversea, and there might well be hostile inhabitants on land (there usually were, in fact), whereas there were none on the sea. One might almost suspect, therefore, that explorers would content themselves with the coastlines and let the interior go.

This was not so. The exploration of the continental interiors, in some cases, followed hard upon the landings on the coastlines. One lure that drove some of the early explorers onward, despite risk and privation, was the hope of finding civilizations whose wealth could be despoiled or human beings, civilized or not, who could be enslaved. This goal was all the more seductive since there were early cases in which it succeeded beyond expectation.

Juan Ponce de León (1460-1521), who dealt in slaves in Puerto Rico, sailed northwestward on March 3, 1513, in search of more slaves. He reached a portion of the North American continent during the Easter season and called it Florida ("flowery") because of its green and flowery appearance. He did not gain much in the matter of the slave trade, however.

More fortunate in this grim business of exploitation and destruction was Hernando Cortés (1485-1547), who in February 1519 sailed from Cuba to Mexico with eleven ships carrying 700 soldiers. In Mexico he found the Aztec civilization with its capital at Tenochtitlán (on the site of what is now Mexico City). Cortés had few men and he faced thousands of strong and brave fighters. Cortés, however, had horses and cannon, which the Aztecs had never seen. What's more, the Aztecs were not united and their resistance was weakened by their early conviction that the Spaniards were gods.

Under the circumstances, the Spaniards found it possible to destroy the Aztec civilization, enslave the Indians, and take their gold and whatever else of value they possessed. There were no qualms of conscience about such things in those days, particularly since the Spaniards had the comfortable feeling that non-Christians had no rights a Christian need respect.

In succeeding years, Cortés and others explored the land, searching for more gold. Cortés himself first sighted the peninsula we now call Baja (Lower) California and the body of water (the Gulf of California) that separates it from Mexico itself.

Cortés's success in Mexico led to a flurry of exploration farther north in a continuing search for more gold. Pánfilo de Narváez (1480–1528) explored the northern shores of the Gulf of Mexico west of Florida and then moved inland. He found no gold and had to struggle back to the coast. There he built five ships and tried to sail across the Gulf of Mexico, but was lost in a storm.

Not all the party was lost, however. Some, under Alvar Nuñez Cabeza de Vaca (1490–1557), were wrecked in what is now Texas. Cabeza de Vaca was imprisoned by Indians for six years, but finally escaped and made his way overland to Mexico City in 1536.

Once he returned, Cabeza de Vaca told colorful stories of his adventures, describing vast herds of buffalo and retailing rumors of great wealth somewhere in the north.

This seemed the more convincing since a second American civilization, that of the Incas of Peru, had been despoiled and destroyed just a few years earlier. Francisco Pizarro (1470–1541) had in 1531 reached Peru and, within two years, had repeated Cortés's feat.

Hernando de Soto (1500–42), who had been second in command under Pizarro, yearned to lead an expedition north of Florida to find there another golden Peru. He landed on Florida's west coast on May 25, 1539, with 500 men and 200 horses, and he struck inland, traveling through what is now the southeastern United States.

On June 18, 1541, he and his men became the first Europeans to set eyes upon the Mississippi River, probably some miles south of the modern city of Memphis, Tennessee. The expedition crossed the river, still heading westward, then turned south. They found no gold but did find plenty of hostile Indians. On May 21, 1542, de Soto died of a fever at a time when they had reached the Mississippi again. De Soto was buried in the river, and the rest of his men made boats, floated downriver, and sailed back to Mexico.

Almost simultaneously, another Spanish expedition was exploring what is now the southwestern United States, under the leadership of Francisco Vásquez de Coronado (1510–54). He too had listened to the tales of Cabeza de Vaca, and between 1540 and 1542 he and the men he led wandered widely over Texas and the Southwest. They were the first Europeans to see the Grand Canyon, but they too found no gold.

The English colonists who settled Virginia and New England in the early 1600s and the Dutch colonists who settled what is now New York explored their respective regions. Their deeds were outdone, however, by the French, who, still at the same time, settled along the St. Lawrence River in the region they called Canada.

Samuel de Champlain (1567–1635) founded Quebec in 1608, discovered Lake Champlain the next year, and, on a later occasion, trekked westward. In 1615, he reached Georgian Bay, the eastward extension of Lake Huron. He was the first European to reach the Great Lakes.

In 1634, Jean Nicolet (1598–1642), a follower of Champlain, crossed Lakes Huron and Michigan and was the first European to reach what is now the American Middle West. Jesuit missionaries followed, intent on converting the Indians.

One of them, Jacques Marquette (1637–75), together with a fur trapper, Louis Jolliet (1645–1700), followed up Nicolet's report of a river west of the Great Lakes. The Great Lakes, after all, could be reached by water from the Gulf of St. Lawrence. If the river in question led to the Pacific, there would be a water route across North America with just a short land interruption between the Great Lakes and the river.

On June 17, 1673, they reached the river, which turned out to be the upper Mississippi, and they followed it downstream for some 1,100 kilometers, by which time it was plain that it was heading for the Gulf of Mexico.

Robert Cavelier de La Salle (1643–87) followed the Mississippi to its mouth in 1682.

In the 1730s and 1740s, Pierre Gaultier de La Vérendrye (1685–1749) pushed westward from Lake Superior and reached Lake Winnipeg and the Black Hills of South Dakota. Two other French explorers, Pierre and Paul Mallet, reached what is now Colorado and caught a glimpse of the Rocky Mountains.

The entire Mississippi Valley was claimed by France as a result of these explorations. Great Britain won the eastern half of the valley in 1763 after the French and Indian War. This passed to the United States when it won its independence in 1783, and in 1803, the western half was bought by the United States.

Thomas Jefferson (1743–1826) was president of the United States at that time, and he planned an exploration of the newly gained region. He entrusted the task to Meriwether Lewis (1774–1809) and William Clark (1770–1838). This "Lewis and Clark expedition" started from St. Louis on May 14, 1804, followed the Missouri River upstream, crossed the Rockies, and moved into the "Oregon Territory," which was not yet under the control of any nation. They then followed the Columbia River to the Pacific Ocean, which they reached on November 15, 1805. They had made the first overland trip across what is now the United States, to the Pacific Ocean and back.

The feat was anticipated by Alexander Mackenzie (1755–1820), however. A Scottish immigrant to Canada, he established himself in what is now Alberta and followed what is now called the Mackenzie River to its mouth in the Arctic Ocean, in 1789. In 1793, he crossed the Rocky Mountains to the Pacific Ocean in what is now British Columbia.

## INSIDE SOUTH AMERICA AND AUSTRALIA

The Spaniards, having settled the northern, western, and southeastern shores of South America, and the Portuguese, having settled the northeastern shores, moved inward by degrees, searching for gold, for slaves, for land, and for Christian converts.

The one great exploratory journey was that of Francisco de Orellana (1490–1546), who had been with Pizarro's band that had conquered Peru. Pizarro set up an expedition to explore the land eastward of the Inca dominions, and Orellana, having reached the headwaters of a great river, felt it would be easier to go ahead than to return over the Andes mountain range, which he had already crossed on his journey eastward.

From April 1541 to August 1542, he progressed down a river which, as it happens, is by far the greatest in the world in terms of water volume. His report mentioned tribes which, it seemed to Orellana, were led by women. This reminded people of the Amazons, the women warriors of Greek legend, and the river was named the Amazon River. Orellana organized a second expedition to the

Amazon, but that proved a disaster. The ships foundered and Orellana himself died near the mouth of the river he had explored.

Despite Orellana's exploratory voyage, Spain could not establish a claim to the Amazon Valley. Portuguese settlers on the Brazilian coast, especially from the São Paulo region, moved inexorably westward, and eventually half of South America, including the entire Amazon region, became part of Brazil.

In Australia, the first sizable settlements were made on the southeast coast in what is now New South Wales, where the town of Sydney was founded in 1788. The first immigrants to the area were convicts transported there as punishment—many of them political prisoners rather than criminals. Voluntary immigrants eventually followed.

Exploration of the interior began with William Charles Wentworth (1793–1872), who in 1813 made his way across the mountain range lying 100 kilometers west of Sydney. Charles Sturt (1795–1869) in 1828 discovered the Darling River (named for Ralph Darling, governor-general of New South Wales, whom Sturt had served as secretary). He followed it to its mouth.

Edward John Eyre (1815–1901) explored the desert areas of southern Australia in the early 1840s but was unable to reach the central regions of the continent any more than Sturt had been.

One of Sturt's companions in exploration was John McDouall Stuart (1818–66). He made six attempts to reach the central region, and in 1860 he succeeded. Then, in his sixth and final journey, he crossed the continent from south to north.

## INSIDE AFRICA

Africa is a study in contrasts. The northeastern corner (Egypt) represents one of the two earliest civilizations in the world. North Africa, generally, was an integral part of the Mediterranean culture and was a part of the Roman Empire first and of the Islamic world thereafter. Nevertheless, all that lay south of Roman North Africa remained unknown to Europeans right down into modern times.

The Sahara Desert is responsible for that. Its broad band of arid emptiness was as much a barrier to European exploration as a great mountain range would have been.

From the time of Henry the Navigator, the seacoast of Africa had come to be known—even earlier than the coasts of the Americas or of Australia. Nevertheless, Africa's interior was the last of the habitable areas to be satisfactorily explored. Its climate, its dangerous animals (whether large mammals or small insects), and its hostile population all combined to make Africa "the dark continent."

The key to the African interior was, of course, the great rivers. There is the Niger River, for example, which enters the Gulf of Guinea on the southern shore of the western bulge of Africa. More than three centuries after its mouth had become well known, its inland course remained shrouded in mystery.

The first European to explore the Niger was Mungo Park (1771–1806). Between 1795 and 1797, he sailed up the Gambia River on the west coast of the western bulge, then worked his way overland to the course of the upper Niger in what is now Mali. He then made his way down the Niger, partly by water and partly by land. In the course of all this, he was imprisoned at one point by Arabs for four months, and nearly died of fever at the end. A second attempt to cover the route of his explorations in 1805 ended in disaster. Park's party was attacked by Africans in what is now Nigeria, and Park was drowned.

Another exploring party, under Scotsman Hugh Clapperton (1788–1827), reached northern Nigeria overland. The expedition started at Tripoli on the Mediterranean in 1822 and pushed southward across the Sahara Desert. Blacks and Arabs had performed this feat often in trading caravans, but this was the first time modern Europeans had done so.

In early 1823, Clapperton's party became the first Europeans to see Lake Chad, on the northeastern corner of Nigeria. It was a lake around which black empires had existed in time past, all remaining unknown to European historians. Northern Nigeria was explored, and Clapperton returned with the tale. Like Park, however, he took part in a second expedition to Nigeria and died there.

Clapperton's companion Richard Lemon Lander (1804–34) continued the work and by 1831 had completely charted the course of the Niger. He, too, died there in the course of a second expedition.

The first European to visit Tombouctou (Timbuktu), the near-legendary black trading center and imperial city of the central Niger (but now much decayed), was the Scottish explorer Alexander Gordon Laing (1793–1826), who reached it on August 18, 1826, after being wounded in a skirmish. He was killed two days after he left the city.

More fortunate was a French explorer, René August Caillié (1799–1838), who reached Tombouctou on April 20, 1828. He did this by studying Islam and learning Arabic so that he could disguise himself as an Arab and join a caravan traveling from Egypt to the city, then joining another that traveled from the city to Morocco.

Involved were numerous hardships, including five months of illness.

During the next half-century, numerous explorers worked out the geographic details of West Africa.

The most intriguing mystery of the African continent was the matter of the source of the Nile. The ancient Egyptians had traveled upstream as much as 2,500 kilometers at their successful imperial height and had found no sign of an end to it.

The first European to ascend the Nile was the English traveler Richard Pococke (1704–65), who worked his way upstream to what is now Aswan in southern Egypt in the late 1730s. In 1770, the Scottish explorer James Bruce (1730–94) traveled upstream on the Nile to Khartoum in the Sudan. There two rivers join, and he moved up the one coming from the east, the one now called the Blue Nile.

This took him into what is now western Ethiopia. He found the source of the Blue Nile in Lake Tana and felt the problem was solved. He was wrong, however. It was the western White Nile, the branch he had not followed, that was the main stream, and its source was still a mystery.

Arab slavers had brought back tales of great lakes in East Africa, and it seemed to some European explorers that those might well be the source of the White Nile (assuming the lakes actually existed).

Richard Francis Burton (1821–90), who later achieved his greatest fame as the translator of *The Thousand and One Nights,* and John Hanning Speke (1827–64) initiated such explorations.

Burton, like Caillié, had studied Arabic and managed (in disguise) to visit the Islamic holy city of Mecca in 1853. He was the first European to smuggle out sketches and measurements of the central shrine of the Ka'abah. He then went to East Africa in 1854 and (again in disguise) entered the holy Ethiopian city of Harar, and was the first European to leave safely after having entered.

In 1857, he and Speke, on their second try to find the source of the Nile, started from Zanzibar on the East African coast, crossed what is now Tanzania, and in February 1858 reached Lake Tanganyika, the long, narrow lake that is on the western border of Tanzania, 1,000 kilometers inland from Africa's eastern coast.

Speke and Burton quarreled at this point. Speke felt that the source of the Nile could be discovered if further lakes were sought out. Burton disagreed, and the two separated. Speke moved northward and on July 30, 1858, reached the largest lake in Africa, which he named Lake Victoria. He felt this would be the source of the Nile, and in a later expedition on July 28, 1862, he located the Nile's emergence from the lake. Counting from the headwaters of the longest stream that flows into Lake Victoria, the Nile River proved to be 6,738 kilometers long, the longest in the world.

The African explorer who most entered the public consciousness was a Scottish missionary named David Livingstone (1813–73). He reached Cape Town in 1841 and pushed northward in an attempt to convert the blacks. (He also argued for their decent treatment by white settlers.) He moved farther north than any European had previously and was the first to explore the Kalahari Desert in what is now Botswana.

Livingstone was intent on pushing far into Africa and opening the continent to civilization and commerce, rather than leaving it to the specialized interests of the slave traders. This meant that he had to face not only the natural perils of the African environment but the sustained hostility of the Boers and the Portuguese. In 1855 and 1856, he wandered across Africa well north of the Kalahari, reaching Luanda on the north Angolan coast and then eastward to the Zambesi River and the Indian Ocean. He was the first European to cross Africa overland in the east-west direction, although he did this in the south where the width was not quite 3,000 kilometers. Along the course of the Zambesi, he discovered a waterfall twice the height of Niagara which he named Victoria Falls.

In a second expedition to Africa, from 1858 to 1864, he explored the Zambesi River area under considerable difficulties. In a third expedition, from 1866 to 1873, he pressed into the lake country, and on March 29, 1871, he reached the upper Congo River, 375 kilometers west of Lake Tanganyika.

By now, though, he had fallen out of sight of Europe and such was his fame that there was considerable anxiety over the possibility of his death.

The *New York Herald,* as a publicity stunt, sent one of its reporters, the Welsh-born Henry Morton Stanley (1841–1904), to Africa to search for Livingstone. Stanley landed in Zanzibar on January 6, 1871, and headed into the interior with a well-equipped caravan. He found Livingstone at Ujiji on the eastern shore of Lake Tanganyika, 1,100 kilometers west of Zanzibar, and greeted him with correct English courtesy, "Dr. Livingstone, I presume?" even though, as the only European within a thousand kilometers, he could be no one else.

Stanley brought needed medicine and supplies, but Livingstone died a year and a half later and Stanley took up the task of exploration.

In 1876, he was in the lake country and circumnavigated Lake

Victoria. He went on to the westernmost point reached by Livingstone and followed the river down to the Atlantic Ocean, thus charting the course of the Congo River.

In his final expedition, he cleared up the remaining uncertainties concerning the lake region and, in 1889, discovered the Ruwenzori Mountain Range, lying on the boundary between Uganda and Zaire with Lake Albert and Lake Edward on either side.

## THE EASTERN ARCTIC

By 1880, the interiors of all the continents lying in the tropic and temperate zones—even Africa—had been roughed out satisfactorily. Only details remained.

The only major regions of Earth that were still unknown to a considerable extent were the polar regions. There the rigors of exploration were even greater than in the African interior.

To be sure, the motivations for the exploration of the polar regions were different in nature from those driving on the investigation of the rest of Earth's surface. In the polar regions there were no civilizations to find, no resources of any value. With the abandonment of any hope for a practical Northeast Passage or Northwest Passage, there was not even the possibility of their playing an important role as a commercial route.

Yet something new had come to the fore, a recrudescence of something very old. This was nothing other than human curiosity, the desire to know for no other reason than to know.

The first Arctic expedition intended solely to increase knowledge was sponsored by Russia. That nation had spread over all of Siberia, and her dominion stretched northward to the ice-covered wastes. Russia wanted to know the extent of her own dominions. In particular, there was the question of whether Asia at its extreme eastern point connected with North America by a land bridge. Dezhnev, who had reached that extreme eastern point, had said there was no land bridge, but was he correct?

The Russian czar Peter the Great, in his last year of life, commissioned a Danish mariner in the Russian service, Vitus Jonassen Bering (1681-1741), to undertake the task. In 1724, Bering traveled overland for 8,000 kilometers, from the Russian capital of St. Petersburg on the Baltic Sea to Kamchatka, a large peninsula jutting southward from far-eastern Siberia. There he built ships and began a sea exploration, following the coast of Kamchatka northward and reaching the eastern tip of Siberia. He found that Dezhnev was right and that there was water between Asia and North America, a waterway now called Bering Strait.

In later expeditions, Bering charted portions of Siberia's northern shore and also explored the sea to the south of Bering Strait, a stretch of water now known as the Bering Sea. He discovered the Aleutian Islands, which swing across the southern border of the Bering Sea in a great arc.

Bering also landed on the North American continent in 1741 and was the first European to reach what is now Alaska. On the basis of Bering's discoveries, Russia claimed and occupied the Aleutian Islands and Alaska, which she held until she sold them to the United States in 1867. The Russians explored Alaska during their period of occupation, reaching Point Barrow, which is Alaska's northernmost point. Point Barrow at 71.26 degrees north latitude is just a little more northerly than North Cape, the northernmost point of Europe, which is at 71.17 degrees N.

The Russians also explored southward as far as the site of San Francisco, which the Spaniards reached from the south, and so between them the two nations completed the mapping of the western coast of the Americas.

Others beside Bering explored the Siberian coast. Under the leadership of S. Chelyuskin, one Russian expedition finally managed, in 1743, to go around the Taimyr Peninsula, which juts northward out of the center of the Siberian shore and is icebound throughout the year. Ships couldn't round the Taimyr through the ice, but Chelyuskin rounded it by sled, and the northernmost point of the peninsula is called Cape Chelyuskin in his honor.

Cape Chelyuskin is at 77.75 degrees N and is considerably farther north than either Point Barrow in Alaska or North Cape in Europe. It is only 900 kilometers south of the North Pole and is the most northerly bit of continental area in the world.

By then it seemed reasonable to suppose that none of the continents extended to the pole, but there might be islands farther north. It remained possible, therefore, that the North Pole might be situated on land.

Some islands were indeed discovered north of the Siberian shore in the century and a half that followed Chelyuskin's feat. Of these, the most northerly bit of land is Rudolfa Island, in the Arctic Ocean 1,960 kilometers north of Arkhangelsk. It is about 75 kilometers farther north than the northernmost shore of Spitzbergen and is only 550 kilometers from the North Pole.

Rudolfa Island is part of a group of islands discovered by the Austrian explorers Julius von Payer (1842–1915) and Karl Weyprecht (1838–81) in 1873. They named it Franz Josef Land after the Austrian emperor. The group comprises the most northerly bits of land in the Eastern Hemisphere, although, of course, in 1873 no one could yet be quite sure of that.

## THE WESTERN ARCTIC

There was the possibility that land stretched farther north in the Western Hemisphere than in the Eastern. The islands that make up a large archipelago north of Canada and, above all, the gigantic island of Greenland looked hopeful in this respect.

The British government, which controlled Canada, was anxious to know the details of its Arctic dominions. Two explorers, John Ross (1777–1856) and William Edward Parry (1790–1855), together and singly, headed expeditions that butted their ways among the islands of the Canadian Archipelago. Between them, they sighted nearly all the major islands by 1833, but had not quite gotten through the archipelago into the ice-choked sea that would lead to Bering Strait and Asia.

In 1833, Ross explored the shores of the Gulf of Boothia (named for Felix Booth, a liquor merchant who had financed the expedition). The western shore was the Boothia Peninsula, which, as it turned out, represents the northernmost stretch of North America. Its extreme point (discovered in 1847 by the Scottish explorer John Rae, 1813–93) is at 71.74 degrees N, about 60 kilometers farther north than Point Barrow, but falling 265 kilometers short of the mark of Cape Chelyuskin.

The next attempt was made by the English explorer, John Franklin (1786–1847). Exploring by land at first, he had mapped almost the entire Arctic coast of Canada between 1820 and 1825. In 1845, he took up the task by sea. He had technology on his side. He was using steamships, not sailing vessels, and the cabins were centrally heated. In the course of the exploration, however, he, his men, and his vessels disappeared.

Other expeditions went in search of them—forty expeditions in fourteen years. Relics of the expedition were discovered in 1853. In the course of these searches, the remaining bits of the Canadian Archipelago were put into place.

In the course of the search, a British naval officer, Robert John McClure (1807–1873), went through the Bering Strait northward in 1850, then sailed eastward. He came upon the westernmost island of the archipelago, where he was forced to abandon his ship. However,

McClure's voyage, added to what had been done in the east, completed the tracing out of the Northwest Passage. No one passed through it in its entirety, however, till the Norwegian explorer Roald Amundsen (1872-1928) accomplished the feat in leisurely fashion between 1903 and 1906.

The northernmost island of the Canadian Archipelago is Ellesmere Island, named in 1852 for a member of the British Parliament, Francis Egerton, Earl of Ellesmere. It is separated from the northwestern coast of Greenland by a narrow stretch of water only 10 kilometers wide in places.

An American explorer, Elisha Kent Kane (1820-57), was the first to approach that narrow strait in 1855. He reached a point at 80.6 degrees N before ice prevented his further progress. This was almost at the level of the northernmost land points of the Eastern Hemisphere, and yet both Ellesmere Island and Greenland seemed to continue onward far to the north.

In 1871, another American explorer, Charles Francis Hall (1821-71), managed to work his way between Ellesmere Island and Greenland to 82.2 degrees N. This was only 550 kilometers from the North Pole, and there was still land farther north. It was not till 1907 that a Danish explorer, Ludvig Mylius-Erichsen (1872-1907), worked his way that far north up the eastern (more frigid) coast of Greenland. He died in the course of the expedition.

Meanwhile, explorers were beginning to penetrate the Greenland interior.

The first significant incursion from the coast was made in 1878 when the Danish explorer Jens A. D. Jensen (1849-1936) moved 70 kilometers inland, reaching an icy height of 1.5 kilometers above sea level. It was clear that the entire interior of Greenland, an area three times that of Texas, must be covered by a thick ice sheet, and all later explorations confirmed this.

In 1888, the Norwegian explorer Fridtjof Nansen (1861-1930) finally managed to cross Greenland east to west on snowshoes and skis. He did this at a line somewhat south of the Arctic Circle where Greenland is about 520 kilometers wide. As was to be expected, he found the ice sheet unbroken, and at one point on the trip he found himself 2.7 kilometers above sea level.

In 1892, the American explorer Robert Edwin Peary (1856-1920) explored the Greenland ice sheet northward and found the limit of its extension at about 82 degrees N. Farther northward was a bare and barren stretch of land that is now called Peary Land, and with that, the northernmost extensions of Ellesmere Island and Greenland were soon worked out. Neither extended to the North Pole.

The northernmost point of Ellesmere Island (Cape Columbia) is at 83 degrees N, which is 235 kilometers closer to the North Pole than Cape Chelyuskin is. As for Greenland, its northernmost point, at the northern end of Peary Land, is Cape Morris Jesup (named for an American banker who financed Arctic expeditions). Cape Morris Jesup is at 83.63 degrees N, about 25 kilometers farther north than Cape Columbia.

As it turned out, Cape Morris Jesup has the distinction of being the northernmost piece of land in the world. It lies 475 kilometers from the North Pole.

At first, of course, no one could be sure there was no land farther north than Cape Morris Jesup. There might be sizable islands at any point up to the North Pole itself, and there was some scientific interest in finding out. In addition, something that was almost a mania drove them on—a lust for record-holding. The question had arisen as to who was to be first to reach the North Pole and to win the honor and the immortality that went along with that.

Ever since the ancient Greeks had first deduced that Earth must be spherical, it was realized that all northward directions must end in a point—that is, that Earth rotates about an axis that stretches from that northernmost point through the center of Earth to a southernmost point. Those intersections of the line of the axis with the surface of Earth are the North Pole and the South Pole.

These points represented an extreme, and were, therefore, noteworthy in themselves, but also, the ice and murderous weather that guard them made the achievement one that could be carried out only at enormous peril to life. This made the task all the more tempting to some.

The first to make a serious attempt to reach the North Pole was Nansen, who had safely crossed Greenland. Ships which had been inadvertently frozen into the polar ice had drifted long distances with the ice before being released in the spring. It occurred to Nansen that a ship specially constructed to withstand being crushed by the ice might deliberately let itself be frozen in and that it might then be carried over the North Pole.

In 1893, he carried through this plan, but found that the drift was not across the North Pole. The ship was never carried much farther north than 85 degrees N. In the course of the drift Nansen left the ship and pushed northward with dogsleds to 86.22 degrees N, managing to come within 420 kilometers of the North Pole, farther north than any bit of land on Earth.

Nansen measured the depth of the Arctic Ocean under its ice

cover. It was deep and seemed to be getting deeper as one went farther north. This lowered the chance of finding any far-northern land, but that did not affect the race for the North Pole.

Peary, the discoverer of the northernmost extension of Greenland, was particularly active in the race. He used the northern reaches of Ellesmere Island as his base, since it was the farthest north one could go by surface vessel through open water and would leave him the shortest distance, less than 800 kilometers, for sledging over ice.

In 1905, his first major attempt from Ellesmere Island brought him to 87.10 degrees N, just 320 kilometers from his goal, before he was forced to turn back.

In late February 1909, Peary went all out. He started with a large party including 24 men, 133 dogs, and 19 sledges carrying 3 1/4 tons of supplies. Depots were established en route, with some of the party dropping off at each depot. Finally, Peary, his dogsled driver, Matthew Hensen (a black), three Eskimos, and some dogs made the final dash from the northernmost depot. They reached the North Pole at last on April 6, 1909. They then returned, following their own tracks and making use of the depots built on the trip northward. By April 25, they were back on the ship.

The increasing success of Arctic exploration was due to the fact that explorers learned to dress and live Eskimo-style and to use dog-power rather than manpower in pulling sledges.


## THE ANTARCTIC WATERS

The South Polar region, the Antarctic, was an even harder knot to unravel than the far north. The Antarctic is much farther from the centers of exploratory activity in Europe than the Arctic is, for one thing. And, as it turned out, the Antarctic is colder and more hostile than the Arctic.

The first serious approaches to the Antarctic on the part of European navigators were the reach to the southernmost part of Africa by Dias and by da Gama, and the passage through the Strait of Magellan by Magellan. In both cases, there was no interest at all in anything that lay farther south. The navigators were merely trying to reach the Indies.

In passing through the strait, Magellan reached a point at 53.92 degrees S, well beyond any southern point ever attained by any European before that. This, as it happens, is the farthest south of any inhabited continental area. To the southeast of the Strait of

Magellan, however, was a land occupied by human beings, for watchfires were seen on it. Magellan called it Tierra del Fuego ("Land of Fire"), a name it bears to this day.

Geographers suspected at first that Tierra del Fuego was part of a great southern continent, for no rational reason other than that the ancient Greeks had speculated that such a continent might exist.

In 1578, however, the English navigator Francis Drake (1540–96) had passed through the Strait of Magellan on his way to plunder the Spanish settlements on the west coast of South America. On emerging into the Pacific, he was struck by a storm that sent him reeling southward far enough for him to see that south of Tierra del Fuego lay open ocean, a stretch of water called Drake Strait ever since.

The southernmost point of land on or near Tierra del Fuego is Cape Horn at 56 degrees S, about 3,850 kilometers from the South Pole.

No one was interested in going farther south than that, at least at that time, but during the explorations of the South Pacific that led to the discovery of Australia, some far-southern bits of land were discovered.

In 1738, a French navigator, Pierre Bouvet de Lozier (1705–86), came across what is now called Bouvet Island in the South Atlantic. It was an uninhabited dot of land at a latitude of 54.43 degrees S. Another French navigator, Yves Joseph de Kerguélen-Trémarec (1734–97), discovered Kerguelen Island in 1771, a cluster of about 300 islets in the southern Indian Ocean at 49.5 degrees S.

Neither island was as far south as Cape Horn, but both were bleak and desolate, a sign of the greater cold of Antarctic areas, since similar land areas at similar latitudes in the northern hemisphere were more inviting.

During the course of his second voyage, Captain Cook pushed so far southward that he finally crossed the Antarctic Circle. He and his crew were the first people in history—not merely the first Europeans, but, as far as we know, the first human beings of any kind—to make the crossing. The date was January 17, 1773. Cook made two later crossings in the course of the voyage, and his most southerly penetration took place on January 30, 1774, when he reached 71.17 degrees S, at which time he was only 1,820 kilometers from the South Pole. Throughout his voyage he was stopped by ice and never saw any actual land.

In the course of his voyage, Cook discovered islands lying to the east and southeast of Cape Horn. The southernmost of the small islands of this group is Thule Island at 59.43 degrees S, 3,460 kilometers from the South Pole.

One consequence of Captain Cook's explorations was the discovery that the Antarctic waters were rich in seals and whales. That drew ships southward where the pure joys of exploration might not have sufficed. A British sealer, William Smith, discovered the South Shetland Islands in 1819. These were due south of Cape Horn, and the southernmost bit of land among them is at 63 degrees S, 3,050 kilometers from the South Pole.

This did not remain a record long. On November 16, 1820, a twenty-one-year-old American sealer, Nathaniel Brown Palmer, sighted land south of the South Shetland Islands. He may not have been the first to do so. William Smith may have preceded him, and a British naval commander, Edward Bransfield, may also have done so. The waters south of the South Shetlands are called Bransfield Strait.

The nature of the land thus sighted was not at first understood. Eventually, it was found to be a peninsula and is now called the Antarctic Peninsula. It is indeed the northernmost extension of a continent. The land sighted in 1820, however, was still well north of the Antarctic Circle.

In that same year, a Russian explorer, Fabian Gottlieb Bellingshausen (1778–1852), circumnavigating the waters in the vicinity of the Antarctic Circle, discovered a small island he named Peter I Island. It was at 68.8 degrees S, which is 240 kilometers south of the Antarctic Circle. It was the first piece of truly Antarctic land ever discovered.

Bellingshausen also discovered a much larger piece of land just west of the base of the Antarctic Peninsula, and this lay farther south still. He called it Alexander I Island. The water surrounding these islands is Bellingshausen Sea.

An English whaler, James Weddell (1787–1834), found a stretch of ocean that extended farther south than anything explored before and on February 20, 1823, reached a mark of 72.25 degrees S before winds and ice turned him back. This represented a new southward record that outdid Cook's fifty-year-old mark. Weddell had approached to a bit less than 1,800 kilometers of the South Pole. The inlet into which Weddell had sailed lies east of the Antarctic Peninsula and is now known as the Weddell Sea.

All the land discoveries of the 1820s in Antarctic waters were made in the general area south of Tierra del Fuego. In 1831 came the first sighting of Antarctic land on the other side of the world. In that year, the English navigator John Briscoe saw a shoreline just north of the Antarctic Circle and south of Madagascar. He called it Enderby Land after the owners of his vessel. He saw it only from a distance, for ice prevented him from actually reaching it.

In 1840, the French explorer Jules Dumont d'Urville (1790–1842) sailed south from Australia and spied a shoreline almost exactly upon the Antarctic Circle, and named it Adélie Land after his wife.

At almost the same time, an American explorer, Charles Wilkes (1798–1877), was following a long stretch of coastline between Enderby Land and Adélie Land, a stretch that followed the Antarctic Circle with surprising exactness. This stretch of land, lying south of the Indian Ocean, is now known as Wilkes Land.

## ANTARCTICA

Wilkes, on returning, was the first to proclaim that all the isolated discoveries of the previous twenty years could be fitted together to indicate the existence of a South Polar landmass of continental size. This marks the first realization that there existed on Earth a seventh continent, nearly twice the size of Australia, but almost entirely within the Antarctic Circle and therefore uninhabited and (except for specially equipped exploring and scientific parties) uninhabitable. The continent was named Antarctica.

In January 1841, the Scottish explorer James Clark Ross (1800–62), a nephew of the Arctic explorer John Ross, entered an inlet into Antarctica in a region that lay generally south of New Zealand, and is now called the Ross Sea. He sailed south till he found himself stopped by a towering wall of ice some 60 to 90 meters high.* This turned out to be an ice shelf, a thick layer of ice extruded over the sea from the vast ice sheet (nine times as large as that which covers Greenland) that existed on Antarctica. The Ross Ice Shelf, as it is called, covers an area of otherwise open ocean that is as large as France.

The Weddell Sea also has an ice shelf over its southern reaches, one that is called the Filchner Ice Shelf after the German explorer Wilhelm Filchner (1877–1957), who first explored it. Neither ice shelf extends to the South Pole. The Ross Ice Shelf, which cuts more deeply into the continent, reaches to 86 degrees S, less than 500 kilometers from the South Pole.

On January 23, 1895, a Norwegian whaler commanded by Leonard Kristenson debarked a party on Victoria Land on the western rim of the Ross Sea. For the first time in the history of the world, human beings stood on land within the Antarctic Circle.

---

* A meter is one-thousandth of a kilometer and is about 1.1 yards long, or 39.37 inches long, in common American units.

One of that party was Carsten E. Borchgrevink (1864–1934), who returned in 1898 and, with nine other men, wintered in Antarctica, the first ever to do so. At one point, Borchgrevink put on skis and set off on the very first attempt to penetrate southward overland. On February 16, 1900, he attained a southern mark of 78.8 degrees S and was only 1,250 kilometers from the South Pole.

Other explorers joined in the competition, one of the most eager being the Englishman Robert Falcon Scott (1868–1912). He and his colleagues sledged over the Ross Ice Shelf and on December 13, 1902, reached 82.28 degrees S, only 800 kilometers from the South Pole.

One of his colleagues, Ernest Henry Shackleton (1874–1922), tried again. On January 9, 1909, his party of four men managed to reach 88.38 degrees S, only 155 kilometers from the South Pole. Each man had dragged his own sledge, and they had turned back only when it was clear that to travel farther would mean their food supply would not last the return journey.

All was set now for the final push. Two candidates were in the field. One was Scott and the other was Amundsen, who had already made his mark in Arctic exploration.

Amundsen prepared with the utmost care, making use of dogsleds and plenty of dogs. There were fifty-two dogs at the start when Amundsen left on October 20, 1911. As he proceeded, he killed the weaker dogs and fed them to the stronger ones, saving the food supplies he had brought for the human members of the expedition. In this way he avoided being forced to turn back as Shackleton had been. Amundsen reached the South Pole on December 14, 1911, and the expedition was back on January 21, 1912, with twelve dogs still surviving and plenty of food left. There had been no human casualties.

Scott's attempt was less carefully organized, did not depend so much on dogs, and was plagued by misfortune. The last 650 kilometers was made by man-hauled sledges only. Scott and four companions reached the South Pole on January 17, 1912, and found Amundsen's marker already there. It had taken them sixty-nine days to reach the pole as compared to Amundsen's fifty-five, and they were worn out. On the voyage back, all five were caught in a nine-day blizzard that was the final straw, and all died of cold on or about March 29, 1912.

By 1912, then, the shape of all the land and sea areas of Earth were worked out in considerable detail as a result of five centuries of more or less nonstop exploration, mostly by Europeans, from the time of Henry the Navigator.

It was only at the end of this period that the recently developed airplane made it possible to pass over all obstacles at great speed and with comparatively little trouble. On May 9, 1926, two Americans, Richard Evelyn Byrd (1888–1957) and Floyd Bennett (1890–1928), flew from Spitzbergen to the North Pole and back in a nonstop flight that lasted only fifteen hours. The first air flight over Antarctica was made on December 20, 1928.

In its essentials, the world was finally mapped.

# 4
# Surface Ups
# and Downs

## MOUNTAINS

The horizon of earthly space did not disappear once the whole surface of Earth had been essentially penetrated and maps (except for comparatively minor details) could be made of the entire globe with reasonable precision.

There was room for expansion even so.

For instance, although Earth's surface is essentially a two-dimensional structure (albeit the curved surface of a sphere), it has its ups and downs, which represent special problems. The prime example of this is the great mountain ranges.

Mountains have always impressed humanity. The great hulks rearing their tops to the sky, misty, silent, snow-covered, and seemingly impregnable, have at once inspired and humbled people. Lifted unreachably high, they have seemed fit habitations for gods rather than human beings, and people have often thought of mountains as sacred ground, and have felt that it made more sense to sacrifice to the gods on hills rather than in the plains.

In the Bible, God thundered from Mt. Sinai, and that was where the Children of Israel, according to the biblical story, received their Law. The Bible doesn't specify the location of the mountain in any identifiable way, but tradition identifies Mt. Sinai with one of the peaks at the southern tip of the Sinai Peninsula, 380 kilometers south of Jerusalem. Its height is 2,285 meters, and to the local Arabs it is Jabal Musa ("Mount of Moses"). The nearby peak of Jabal Katherina is 2,670 meters high and would make a more impressive home for God, but there is no arguing with tradition.

Similarly, in the Book of Deuteronomy, the Israelites are instructed to perform various important rites on mountain peaks within the new land they were to occupy. Mt. Gerizim and Mt. Ebal, in central Israel, were selected for the purpose. In time, the temple built by Solomon in Jerusalem became the center of Judean worship, completely replacing all "the high places," but among the Samaritans, a Jewish sect, Mt. Gerizim remained the holy place. It is only 885 meters high.

The ancient Greeks considered the home of the gods to be on Mt. Olympus in northern Greece. It is 2,900 meters high and is the tallest mountain on the Greek peninsula. Again, Fujiyama, 3,775 meters high, and the highest in Japan, is a sacred mountain.

As time went on, religious notions etherealized and the place of the deities was transferred from an earthly mountain peak to the sky, and then to a transcendent heaven that was not part of the visible universe at all. Nevertheless, it is possible to suspect that the domes and steeples that surmount churches as well as the minarets of Islam, the ziggurats of Babylonia, and the pyramids of Egypt are all examples of a man-made harking back to the mountains where the gods lived.

If we put aside religion and superstition, mountain ranges impinged upon human beings first as barriers. Such barriers could be comforting, since mountains can serve as protection against invading armies, or as places where a grimly determined defense could be put up by a population who knew all the mountain details against invaders who did not.

Thus the Alpine mountain range that curved around the north of Italy protected that land most of the time, whereas those same mountains have always been an important factor in keeping Switzerland free.

Overconfidence in such barriers could do harm, however. Hannibal of Carthage caught the Romans by surprise in 218 B.C. when, instead of attempting to reach Italy by sea from Spain, he took the enormous risk of traveling by land and crossing the Alps.

Usually, though, mountains protect. When the Moslems conquered Spain in A.D. 711, their attempt to add the Frankish realm to their dominions was hampered by the Pyrenees. The Frankish realm held, and when Charlemagne launched a counterattack in 778, the Pyrenees hampered him as well. To this day, the line of the Pyrenees marks the boundary between France and Spain—and that of the Himalayas marks the boundary between India and China, that of the Andes marks the boundary between Chile and Argentina, that of the Alps marks the boundary between Italy and its neighbors, and so on.

## CLIMBING THE ALPS

In ancient and medieval times, no one seems to have climbed mountains for any other reason than to get to the other side. In 1492, however (the year Columbus discovered the New World), a French courtier led a group up Mt. Aiguille (about 2,100 meters high) in southeastern France, 45 kilometers south of Grenoble.

The courtier accomplished the task because his monarch, Charles VIII of France (1470-98), had heard that the mountain was supposed to be unclimbable and wished to test that. Being a king, he could order someone else to undergo the risks and endure the hardships involved.

The Swiss naturalist Konrad Gesner (1516-65) climbed Alpine peaks on a number of occasions and seems to have enjoyed doing so. He wrote voluminously on the different species of plants and animals, and his mountain-climbing had a scientific excuse, since he was searching for rare species that were confined to the mountain heights.

Another example of such climbing took place in North America. There, a New Hampshire colonial, Darby Field, in June 1642, climbed what was later to be called Mt. Washington (1,917 meters high and the highest peak in what is now the northeastern United States). He had difficulty persuading Indian guides to accompany him, for the Indians, not surprisingly, considered it holy ground.

It was not till the 1700s, however, that mountain-climbing became more than an occasional quixotic feat. It was at that time that the science of geology—the study of Earth's crust—began to flourish, and scientists found themselves interested in mountains. They studied their formation, their rocks, and their glaciers. Botanists grew increasingly interested in the plant life peculiar to mountains, and zoologists in the animal life.

Because modern science grew and expanded in Western Europe particularly, it was the mountains of Western Europe that received the first detailed attention from scientists—in particular, it was the Alps, the tallest mountain chain in Western Europe.

The tallest peak in the Alps is Mont Blanc (*blanc* is the French word for "white"; the name refers, of course, to the mountain's year-round snow cover), which is 4,807 meters high, more than twice the height of Mt. Sinai. It is in southern Switzerland, just north of the Italian border, about 100 kilometers southeast of Geneva. It and other peaks nearby had the largest and most spectacular glaciers in the Alps, and they attracted much attention therefore.

In 1760, a twenty-year-old student of physics from Geneva named Horace Bénédict de Saussure (1740-1799) arrived at Mont Blanc. In later life, he was a professor of physics who invented instruments to measure atmospheric humidity and also electrical potential. He was also the first to use the word "geology." In 1760, however, he was just a young man looking at Mont Blanc and thinking what a wonderful feat it would be to climb to the top of it.

De Saussure didn't quite have the nerve to do it himself, but he offered prize money for the first person to succeed. It was twenty-six years before a person tried to win the prize. That person was Michel Gabriel Paccard, a French doctor who in 1786, accompanied by a porter, reached the top of Mont Blanc and collected the prize.

Once he knew it could be done, de Saussure waited no longer. The next year, although he was forty-seven by the time, he too climbed Mont Blanc. With him he carried scientific instruments to measure air pressures, temperatures, and so on. Of course, starting with that first sight of the mountain, he had become an enthusiastic mountaineer and seized every opportunity to study the rock formations of the mountains.

He noted that the strata (or layers of rock) were parallel, but that they were not horizontal or even straight. They curved in dips and loops. At first he assumed that that was just how the rocks had happened to crystallize, until he saw that some of the strata consisted of sand and pebbles not very strongly held together. He realized that they could not have been deposited on end. The strata must have been level and horizontal to begin with and must have folded and crumpled in the course of the formation of mountains out of level ground.

Thus the climbing of mountains proved to have scientific significance, but only a small portion of the climbers thought of that. Mountain-climbing became a sport.

In 1854, an Englishman, Alfred Wills, climbed the Wetterhorn, an Alpine peak 3,708 meters high, while he was on his honeymoon. It was an odd thing to do at such a time, but it is an example of the way in which the risky and arduous sport had caught the imagination of the kind of people who found such challenges irresistible.

More and more daredevils (almost always from the British Isles, for some reason) came to scale the mountain peaks, and the Swiss developed a series of guides, a sort of professional corps of mountaineers, who could cater to the tourists.

Some of the fun was taken out of the Alpine sport by the fact that Mont Blanc, the highest of the peaks, had been the first one climbed and there was thus no chance of setting a new height record. However, height alone is not always a true measure of difficulty.

Consider the Matterhorn, for example, about 20 kilometers east of Mont Blanc. It is 4,478 meters high, 329 meters lower than Mont Blanc, but its peak seems to thrust straight upward and it presents an especially formidable appearance, particularly from the Italian side. It was popularly supposed to be unclimbable.

An English artist, Edward Whymper (1840–1911), came to the Alps in order to paint mountain scenes and was caught up in the fascination of mountaineering. In the early 1860s, he made no fewer than six attempts to climb the Matterhorn from the Italian side and failed. At one time, though, he caught sight of the other side of the mountain, the Swiss side, and it seemed to him the slope was more manageable there.

When he tried a seventh time, therefore, he went by way of the Swiss side, and, on July 14, 1865, at 1:40 P.M., he made it to the top, with six companions. On the way down, however, with all seven descending in single file and attached to each other by a long rope, one person—the least experienced of the climbers—slipped. The idea of the rope was to keep a slip from being fatal. Others on the rope would cling to the mountainside and the one who slipped would then dangle till rescued.

This time it worked in the other direction, however. The man who slipped pulled the next man loose, and the doubled weight pulled a third man loose, then a fourth. If the rope had not broken between the fourth and fifth man, all would have plunged off the mountain. The four died, but Whymper and two Swiss guides were saved. (There were wild rumors that one of the survivors had deliberately cut the rope, but this was never proved.)

This dreadful accident did more than anything else to glamorize mountaineering and make it attractive to those to whom danger was a spice.

As so often happens, once a feat of skill and daring is accomplished with great difficulty after years of attempts, repetitions become common place, almost easy in comparison. Only three days after Whymper's climb, a group of Italian mountaineers reached the top from the more difficult Italian side. Nowadays it is routinely climbed in the summer mountaineering season, and at least 100,000 people have climbed the peak in the last century, although more than 90 have been killed in the process.


## BEYOND THE ALPS

By 1870, the Alps were no longer a real challenge. Too many peaks had been climbed too many times. The Alps, however, were by no means the highest mountain range in the world, and Mont Blanc was far from the highest peak.

Attempts had already been made elsewhere. In the American Rockies, Pikes Peak in Colorado had been climbed in 1819, but it is lower than Mont Blanc.

The first mountain peak higher than Mont Blanc that was successfully ascended was Mt. Ararat, in what is now eastern Turkey, in the corner of the land that lies between Iran and the Soviet Union. It has two peaks, of which the higher, Great Ararat, is 5,165 meters high, 358 meters higher than Mont Blanc.

It was scaled in 1829 by a German, Johann Jacob von Parrot. There was a special motivation behind the climb. Mt. Ararat is, according to tradition, the mountain on which Noah's ark finally came to rest after the Flood, and there is always the hope among the romantic that some relic of that ancient vessel will be located.

Von Parrot found no trace of the ark. There have been other climbs since and occasionally reports of relics are heard, but nothing really convincing has ever turned up.

The highest peak in Mexico is Citlaltépetl, about 70 kilometers west of Vera Cruz and 5,700 meters high. It was first climbed in 1848.

It was after 1870, however, that mountaineers of the world began to seek out the great peaks, wherever they happened to be.

Kilimanjaro is the highest mountain peak in Africa. It is located in Tanzania, just south of the border of Kenya, about 210 kilometers south of Nairobi, and is 5,895 meters high. In 1889, two Germans, Hans Meyer and Ludwig Purtscheller, successfully climbed it.

About ten years later, a British mountaineer, Halford Mackinder, climbed the somewhat lower Mt. Kenya (5,193 meters high and 320 kilometers north of Mt. Kilimanjaro).

Kilimanjaro did not set a record, however. Mountaineers had been tackling the Andes mountain range. One of the highest of the Andean peaks is Chimborazo in Ecuador, 150 kilometers south of Quito and 6,267 meters high. It had been tackled a number of times.

In fact, the great geographer Alexander von Humboldt (1769–1859) had attempted the feat as early as 1802, in the early days of mountaineering, and had managed to attain a height of 5,760 meters, nearly 1,000 meters higher than the top of Mont Blanc. However, since he did not reach the top of the mountain, that feat is usually overlooked.

In 1880, Whymper, the conqueror of the Matterhorn, climbed Mt. Chimborazo, not once but twice. He was the first to climb a peak higher than 6,000 meters.

There are a few Andean peaks higher than Chimborazo. The highest in the Andes mountain range and, indeed, the highest anywhere outside Asia is Mt. Aconcagua, on the border of Chile and Argentina and 100 kilometers northeast of Santiago. Its height is 6,960 meters. It was climbed in 1897 by an expedition headed by an English mountaineer, Edward A. Fitzgerald.

The highest peak in North America is Mt. McKinley. It is in southern Alaska, 250 kilometers southwest of Fairbanks. Its height is only 6,194 meters, well below that of Aconcagua, but Mt. McKinley's position not far south of the Arctic Circle makes it the most nearly polar of the world's great mountains, and this increases the difficulty of the climb. It was not till 1913 that a party of four, under the leadership of a fifty-year-old English-American Episcopal minister, Hudson Stuck (1863–1920), managed to climb the mountain to its peak.

In the twentieth century, however, mountaineers turned increasingly to the Himalayan complex of ranges, which include the highest mountains in the world. There are on Earth some forty peaks that are higher than 7,000 meters, and every last one of them is in the Himalaya Mountains.

The first of these to be climbed was Trisul in northern India, 300 kilometers northeast of Delhi. It is 7,120 meters high and was scaled in 1907.

The tallest peak in the Soviet Union is Communism Peak, 7,495 meters high, in the Pamir Range, the westernmost extension of the Himalayas. It is in the Tadzhik S.S.R., about 120 kilometers west of the Chinese border. A team of Soviet climbers managed to reach its peak in 1933.

That did not set a record, however, for in 1931, Kamet, a peak 70 kilometers northwest of Trisul, had been scaled, and its height is 7,756 meters. Five years afterward, that record was broken when an English party scaled Nanda Devi, 7,817 meters high, in the same region.

The true aristocrats of the mountain ranges were the fourteen peaks that topped 8,000 meters, most of them on the borderland between Nepal and Tibet.

This was Earth's ceiling, and mountaineering here reached a level of difficulty comparable to that of reaching the poles. In one respect, the mountain peaks were far worse. However difficult the environments of the North Pole and South Pole were, it was at least possible to breathe there. On the great mountain peaks of Nepal-Tibet, however, the air was so thin that the smallest effort came at a great price. In the end, oxygen containers were as important to mountaineers as sledge dogs to polar explorers.

The first 8,000-meter peak to be conquered was Annapurna in north-central Nepal. Its height is 8,078 meters and it was finally scaled in 1950 by a French team led by Maurice Herzog.

The world's highest mountain was one that was precisely on the border of Nepal and Tibet, about 150 kilometers from the eastern border of Nepal. It was not till 1952 that its primacy was proved beyond dispute. A survey then showed that its highest peak is over 8,800 meters, and no other peak is that high. The exact height of this peak is now taken as 8,848 meters. It is 1.84 times as high as Mont Blanc and three times as high as Mt. Olympus. The Tibetans called it Chomolungma ("Goddess Mother of the World"), and for once the awe at a local mountain is not misplaced. In 1865, it had been named Mt. Everest, after George Everest, who had been surveyor-general of India from 1823 to 1843.

Attempts to climb Mt. Everest by British mountaineers began in 1920. Among the most ardent of them was George Leigh Mallory (1886–1924), who, when asked why he wanted to climb Mt. Everest, gave the inspired and forever-quoted answer "Because it's there!" This is the exasperating (and glorious) reason that lies behind so many of humanity's triumphs and disasters.

In 1924, Mallory and a companion, Andrew C. Irvine, struggled upward from the highest depot and were seen by those who remained behind to be within 330 meters of the peak, and then the clouds swirled in. They were never seen again. In 1933, the peak was again almost reached but not quite. Altogether, between 1921 and 1938, there were seven attempts to scale Mt. Everest from the Tibetan slopes, and all failed.

Then came World War II in 1939 and the Chinese Communist conquest of Tibet in 1949. By the time mountaineers were ready to move again, Tibet was closed. In 1951 and 1952, three attempts were made from the Nepalese side, and all failed, though again one of them got within 300 meters of the top.

Finally in 1953 the largest expedition of all was planned, one that made full use of advances in technology. Oxygen cylinders were used, as well as insulated shoes and clothing, radio equipment, and so on. Eight camps were established, and the ascent was carried out systematically until, on May 29, 1953, Edmund P. Hillary (1919– ), a New Zealand beekeeper, and a Nepalese guide, Tenzing Norkay (1914–   ), stood on the highest bit of land on Earth.

Since then, the climbing of Mt. Everest has been repeated nearly a score of times, and virtually all the other high mountain peaks have been scaled.

In this direction, too, the horizon appears to have reached its limit.


## CAVES

The surface of Earth leads not only upward, however, but to some extent downward, for the uppermost skin of Earth is not necessarily solid. Hollows are formed, either because molten lava solidifies around an air bubble, or because slightly soluble rocks are gradually leached out by rainwater and springs. The results are caves.

In primitive times, such caves formed handy shelters for human beings. They offered protection against predators and weather and could be easily warmed by bonfires. Many fossils and other traces of Ice Age man have been found in caves, and some of those caves contain paintings that are the earliest evidences of the artistic cravings of human beings.

The longest cave system in the world, as far as is known, is the aptly named Mammoth Cave in Kentucky, about 135 kilometers southwest of Louisville. One can follow the intricately interconnected corridors for an extreme of 232 kilometers, double the length of the second most extensive cave system, Holloch Cave in Switzerland.

Speleology, the study of caves, became an organized study toward the end of the 1800s, and so did the exploration of caves as an exciting sport (sometimes called "spelunking").

As early as April 6, 1841, a man named Antonio Lindner had penetrated a cave near Trieste (then in Austria-Hungary) to a point that was 329 meters below the cave entrance. That was the deepest anyone had penetrated until that time, and the mark stood through the rest of the nineteenth century.

The twentieth century saw a flurry of explorations. It was in France that the deepest (as opposed to the most extensive) caves existed. The deepest of all, as far as we know, is the Gouffre de la

Pierre St. Martin in the western Pyrenees, and on November 8 to 11, 1969, an exploring party made it to a point 1,173 meters below the entrance.

We cannot be certain that there are no caves anywhere that are not deeper still, but it is not likely that caves can be very much deeper, for the pressures as one penetrates downward through the rock layers increase rapidly. The temperature also increases rapidly, and the combination of temperature and pressure will compact any cavities out of existence at depths of considerably more than a kilometer. Caves are therefore strictly surface phenomena of Earth, and notions (often found in pseudoscientific fables, in primitive science fiction, and in bad motion pictures) that there may be caverns leading downward to a hollow center of the planet are strictly fantasy.

To be sure, human beings have dug into the ground from earliest times in search of water or metal ores, and they have on occasion dug desperately deep in that search.

In Montana, a water well was dug in October–November 1961 that reached downward 2,230 meters, twice as deep as the deepest known point in a natural cave.

Human beings in search of gold will dig deeper still. South Africa has the deepest mines, and one of them, at Carltonville, reached a depth of 3,540 meters in April 1974. At that depth the temperatures are such that refrigerated ventilation must be used if miners are to be able to work. At that depth, too, the pressures exerted on the rock by the weight of the rock layers above is such that there is the constant danger of the tunnels collapsing. It is not likely that human beings will penetrate much deeper without some important advances in technology.

Even deeper holes are dug for that most desired thing—oil. The deepest oil well presently existing is in Oklahoma, where a depth of 9,583 meters, deeper than Mt. Everest is high, was attained.

# 5
# The Ocean

## THE WATER SURFACE

So far, we have dealt with the expansion of the human horizon chiefly in connection with the land surface. The sea we have considered merely as a highway, as a route from one piece of land to another. Yet surely the sea, which covers 70 percent of Earth's surface, itself offers a field for study—a new horizon.

Human beings have sailed and steamed all over the ocean but have usually interested themselves only in the surface. They have for the most part ignored the huge volume of water that lies beneath the surface—the domain of fish and of many other forms of life, far richer in quantity and far older in evolutionary background than the life on land.

To be sure, human beings are not utter strangers to the world of water. *Homo sapiens* is a thoroughgoing land mammal and can easily remain away from water (except for drinking) throughout life. Nevertheless, human beings can, if they wish, immerse themselves in water and move about in it.

Many land mammals can swim easily, continuing in water the same basic limb movements that make it possible for them to progress over land. Human beings are not that fortunate. Their method of land progression is a highly specialized bipedal one which won't work in water. In water, human beings must revert to the quadrupedal stance and use all four limbs. It is something that must be learned.

Nevertheless, once learned, swimming is a pleasurable activity, and many people engage in it whenever they can.

One can swim while totally submerged. A person can paddle under the surface, or dive under from a height. The human being is not, however, by any means properly adapted for the purpose. People can stay underwater only as long as they can hold their breath, and that, even with practice, is a very limited period of time.

For the average human being, remaining under water for as long as a minute is a difficult chore. Experts, after considerable practice, can achieve a length of perhaps two and a half minutes. For anyone, immersion as long as five minutes is fatal. The brain has by then consumed the available supply of oxygen and is irreversibly damaged.

It is possible for land animals to be far better adapted to water life, of course. The land animals best adapted for the purpose are the whales and their relatives. So adapted are they that although they have evolved from land-animal ancestors they can no longer live out of water. The whales carry their land-animal ancestry in much of their form and functioning; for instance, they have lungs and breathe air even though they are never out of water. They are not condemned to the surface as we are, however. The sperm whale, the largest of the toothed whales, can remain underwater up to an hour and a quarter, while the much smaller bottle-nosed whale has been reported to remain underwater up to two hours.

A human being cannot go very deep, since water pressure mounts rapidly with depth and quickly sets a limit to human endurance. Where the rewards are great, however, human beings press those limits. Pearl divers, who harvest the oysters on shallow sea bottoms, can force themselves down 15 meters or more and remain underwater for two minutes before breaking the surface for additional air. What's more, they were doing so in the earliest days of civilization in the Mediterranean Sea and in the Persian Gulf. Modern records have seen unprotected humans swim down to a depth of as much as 30 meters—but the sperm whale can dive as deeply as 900 meters in its search for food.

Human beings can improve their underwater performance if they use their ingenuity. An airtight transparent cover over the face, while not supplying any air, at least protects the eyes and makes seeing easier. Transparent shields of thin tortoiseshell, very primitive by modern standards, but usable, were devised as long ago as the 1300s by divers in the Persian Gulf.

Glass goggles came into use as early as the 1860s. In the 1930s, short pipes, one end in the mouth and the other sticking up above the surface of the water (a so-called snorkel, from a German word for "snout"), came into use, as a way of breathing while underwater. This did not permit very great depths, however, for the water pressure on the outside of the chest was greater than the air pressure inside the chest, and even at shallow depths it became impossible to expand the chest and inhale against the push of water.

In 1933 rubber foot fins came into use. Goggles, snorkels, and fins made it possible for swimmers to move underwater with greater efficiency and for longer periods of time than through simple unequipped holding of breath. Nevertheless, the swimming could only be immediately beneath the surface. The expression "skin diving" used for such practices indicates that it takes place only just beneath the "skin" of the water.

During World War II divers with face masks and fins were used to reconnoiter shorelines and to attach explosives to vessels beneath the waterline. Such "frogmen" were used first by the Italian armed forces and then by others.

What was really needed, however, was to have an underwater swimmer become independent of surface air, at least for a time. During World War II, skin divers began carrying cylinders of compressed air, which could be exhaled into canisters of chemicals that absorbed the carbon dioxide and rendered exhaled air fit to breathe again. Such systems were first developed in 1943 by the French naval officer Jacques-Yves Cousteau (1910–    ). These were called "aqua-lungs," and the sport, which became very popular after the war, was called "scuba diving," the word "scuba" being an acronym of "self-contained underwater breathing apparatus."

Experienced scuba divers can easily attain depths of 60 meters and remain there for some time. At greater water pressures, however, nitrogen gas dissolves in the blood to a greater extent than at ordinary air pressures, and this can eventually produce a kind of intoxication which lures the scuba diver into going deeper or staying under too long, and death may result.

Yet even a depth of 60 meters is very shallow compared to the total depth of the ocean, and the best scuba diver is still only in the skin of the sea.


## THE WATER DEEPS

The first expression of curiosity about the greater depths of the ocean that we know of was that of the Greek philosopher Posidonius. About 100 B.C., he is supposed to have measured the depth of the Mediterranean Sea just off the shores of the island of Sardinia and to have come up with a depth of 1.8 kilometers by modern measure.

That remained a lonely attempt, if, indeed, it is anything more than a story. It was not until the 1700s that scientists grew interested in the depths of the sea, and then it was primarily out of an interest in sea life.

How could they observe the life forms beneath the ocean surface? They could scarcely accomplish much by diving a short distance and looking about for a short time. Why not reverse things and bring living organisms from the depths to the surface?

In the 1770s, a Danish biologist, Otto F. Muller (1730–84), devised a dredge. It was a strong net attached to an iron frame and could be trailed many meters beneath the water surface. Living things would be tangled in its meshes and would be brought up.

One person who used a dredge with particular success was an English biologist, Edward Forbes, Jr. (1815–54). During the 1830s, Forbes dredged up sea life from the North Sea and from other waters around the British Isles. Then, in 1841, he joined a naval ship that was going to the eastern Mediterranean, where he did more and better dredging than anyone had yet done, studying all the kinds of living things he brought up. He dredged up a starfish from a depth of 400 meters, for instance.

Plant life can live only in the uppermost layer of the ocean, since sunlight does not penetrate more than 75 meters or so. Animal life cannot live except (ultimately) upon plant life. It seemed to Forbes, therefore, that animal life could not long remain below the level at which plants were to be found. In fact, he felt the 400 meters from which he had dredged his starfish was about the limit of life in the sea and that at levels below 400 meters the ocean was barren and lifeless.

Yet even as he decided this, evidence to the contrary appeared.

James Clark Ross, who was exploring the shores of Antarctica in 1841 (see the previous chapter), was intent on doing more than merely plot the shape of those shores. He was trying to find out everything about the ocean off those shores, as well.

If we ignore the dubious story about Posidonius, Ross was the first to try to determine how deep the ocean might be. He let down a long, weighted cable, hoping to hit bottom. He also used dredges more deeply than anyone had before and brought up all sorts of sea life from as deep as 730 meters, well below Forbes's limit.

Ross's report made little impact at the time, but something else happened just a few years later that was seemingly utterly unrelated but threw the investigation of the deep sea into high gear.

In 1844, an American inventor, Samuel F. B. Morse (1791–1872), had constructed the first telegraph line. It ran from Baltimore, Maryland, to Washington, D.C., a distance of 65 kilometers, and for the first time, communication became essentially instantaneous.

The telegraph was a simple invention (once it was made) and could spread its tentacles outward with comparatively little investment. The industrial world of Europe and North America was quickly interconnected by skeins of wires strung along poles rooted in the ground.

Some places were separated by water, however, and telephone poles could not very well be planted over rivers or over arms of the ocean. One way out was to wrap the wires in waterproof coating and make "cables" out of them. The cables could then be laid along the bottom of a stretch of water.

Cables were laid across the bottoms of the Hudson River and the Mississippi River in the 1840s, for instance. In the 1850s, cables were laid across the English Channel and the Irish Sea. That connected England, by telegraph, with Ireland and France.

The big job, however, was to stretch a cable across nearly 5,000 kilometers of the Atlantic Ocean, in order to connect Europe and North America.

This could be very important. In December 1814, Great Britain and the United States had signed a treaty of peace in Ghent, Belgium, that ended the War of 1812. News of the treaty, however, couldn't reach the United States until a ship traveled across the Atlantic Ocean with the information—and that took six weeks. Before the ship reached the United States, the Battle of New Orleans had been fought on January 8, 1815. It was the largest and bloodiest battle of the war, but it was fought when the war was over. Once a cable was laid across the Atlantic Ocean, nothing like that could ever happen again.

An American financier, Cyrus West Field (1819–92), was determined to lay a cable across the Atlantic Ocean. In 1854, he founded an American company devoted to raising the money for the task and to carrying through the necessary work. A British scientist, William Thomson, Lord Kelvin (1824–1907), concerned himself with the scientific aspects of the feat, devising ways to strengthen the signal periodically so that it could survive the transoceanic transmission without being reduced to a mere whispered garble of noise.

A cable was finally laid by 1858, going from Newfoundland to

Ireland, and President Buchanan and Queen Victoria were able to exchange greetings. However, the insulation soon failed and the cable was useless.

The indomitable Field, despite this, and despite the loss of his fortune (which he had built up in the paper business), began all over again. This time he had the help of the *Great Eastern,* the first steamship that could be considered a great ocean liner in the modern sense. It was 211 meters long, far longer and larger than any ship in existence in 1858 when she was launched. She was, in fact, ahead of her time and could never make a transatlantic crossing with a full cargo. She was a technical marvel, but an economic disaster. She finally found the job, however, that she was just right for—the carrying of the tons upon tons of cable to be strung along the ocean floor.

In 1866, the first permanent cable was laid. Others followed, until finally neither the Atlantic Ocean nor any other offered any barrier to electric communication.

All this had its effect on scientific knowledge of the ocean depths. If the cable was going to be laid on the sea bottom, there had to be some knowledge of how far down the sea bottom was, how even or uneven it might be, and whether one route might be better than another.

The man for the job was the American naval officer Matthew Fontaine Maury (1806–73). Retired after a stagecoach accident in 1839 and forced into a desk job, he began a study of ocean winds and currents that made him the "father of oceanography."

In 1850, his part of the great project of laying the cable was the determination of ocean depths and the preparation of a chart that would give a profile of the sea bottom. In those days, the only way of measuring the depth of water at sea was to pay out a long line until it struck bottom and then determine how much had been paid out.

It was not an easy task. The ocean was deep enough to make the required length of line heavy and hard to handle. What's more, it took a great time to pay out and to heave up again, and one could never be certain that it was truly vertical—ocean currents might make it hang obliquely to show a deeper-than-actual depth. Then, too, there was always the chance that the line had sunk into a gully considerably lower than the main ocean floor all about or struck the top of a hill considerably higher than that floor. The general feeling was that the bottom of the ocean was, on the whole, featureless. The movement of the water, it was felt, would tend to scour out unevenness, and the deposits of sediment would fill in gullies. The comparatively few measurements that Maury was capable of supervising did not, and could not, do much to disturb this feeling.

Nevertheless, by 1855, Maury had found that the Atlantic Ocean seemed to be distinctly shallower in the center than on either side. This central shallow region Maury named Telegraph Plateau in honor of the cable that was to carry telegraph messages across the ocean.

One ship that labored to make depth measurements (or "soundings") was the British ship *Bulldog,* which set sail in 1860. On board the ship was a British doctor, George C. Wallich (1815–99), who was in charge of any discoveries that might be made about sea life. He was watching in October when a line was heaved overboard at a spot in the Atlantic Ocean about halfway between the northern tip of Scotland and the southern tip of Greenland.

The sounding line went down to a depth of 2,300 meters, and when it came up, thirteen starfish were found near the lower end of it. What's more, they weren't starfish which had died and sunk to the sea bottom. They were very much alive.

Wallich reported this at once and insisted that animal life could exist in the cold darkness of the deep sea, even without plants.

Biologists were still reluctant to believe this, and a Scottish biologist, Charles W. Thomson (1830–82), took up the matter. In 1868, he went out into the North Atlantic on a ship called *Lightning.*

He then did indeed settle the matter. Dredging through deep waters, he obtained animals of all kinds, and all argument ended. Forbes's idea of a lower limit of life vanished.

Thomson made a particularly important discovery when he measured the temperature of water at different depths. Till then, it had been thought that water in the deep sea was at a uniform 4°C., at which temperature water was at its densest and would naturally seek bottom. Thomson showed, however, that water at a given depth varied in temperature in different places. In some places, it was quite a bit warmer than 4°C.

Where did the water come from? It seemed likely it came from upper layers which were warmed by the sun. This meant there must be water currents that carried water from the surface of the ocean down into the depths. There must therefore be other currents that carried water from the depths to the surface. In other words, water circulated through the whole ocean, and with this, the notion of life in the ocean deeps began to make sense.

The water in the ocean surface dissolves oxygen from the air. Animal life in the sea lives on that dissolved oxygen. The ocean currents carry that dissolved oxygen all the way down to the very bottom, so life can get the oxygen it needs at all depths.

Animals in the deep can get food, too, even in the absence of plants. When an animal eats a plant or animal, bits of the object being eaten may break off and drift downward. Sometimes plants or animals in the surface layers may die without being eaten and drift downward. As this once-living material drifts downward, it is seized and eaten by animals at lower levels in the water, which themselves contribute to the drizzle. Finally the drizzle, added to and subtracted from by animals at all levels, reaches the bottom of the ocean.

If that were all, then all the chemicals in the surface layers of water which support life would gradually be transferred to the ocean bottom. The surface layers would be depleted and life in the surface would die out and, with it, life in the deeps and life on land. However, the ocean currents that carry oxygen from the surface to the sea bottom also, in returning, carry chemicals from the sea bottom to the surface, so that life continues.

In 1869, Thomson went out in another ship, the *Porcupine,* and managed to dredge down to a distance of 4,400 meters, and he still came up with different species of animals. He was certain that life existed down to the bottom of the ocean, however deep that might be. He was determined now on a full-fledged expedition that was to visit all the oceans of the world.

Thomson set out on December 7, 1872, in a ship called *Challenger* and remained at sea for three and a half years. The *Challenger* sailed over all the oceans, for a distance of 125,000 kilometers altogether. The depth of the ocean was measured in 372 different places. In the Pacific Ocean, which turned out to be the deepest as well as the largest, there were places in which the bottom was 7 1/4 kilometers down from the surface. And even from the deepest part of the ocean, living things were brought up.

## THE SEA BOTTOM

Even the *Challenger*'s monumental achievement did not do more than produce the merest outline of the geography of the sea bottom. The 372 soundings represented one sounding per 1,000,000 square kilometers of ocean. To get an idea of what this could do toward the determination of the configuration of the sea bottom, imagine what kind of relief map of the United States there could be if geographers had the figures for the height of seven randomly chosen spots within the nation and nothing more to go on.

Nor could anything more be expected as long as the state of the art remained what it was. Judging depths by allowing weighted lines to drop into the sea was too long, too difficult, too clumsy a task. It could never suffice. Fortunately, the state of the art did not remain what it was.

Sound travels at the great speed of 331 meters per second or 1,192 kilometers per hour under standard conditions. What's more, sound can, under certain circumstances, be reflected, as we learn every time we hear the echo of a shout sent back by a cliffside. A sound wave can conceivably travel three kilometers and be reflected, and the echo can then be detected eighteen seconds later. Indeed, from the time between the original beam emission and the echo detection, the distance of a reflecting object can be determined.

The shorter the wavelength of sound, the more efficiently it is reflected. The best results are achieved by sound waves with wavelengths too short for the human ear to hear ("ultrasonic" sound).

In 1880, a French chemist, Pierre Curie (1859-1906), with his brother, Jacques, worked out a convenient way of producing ultrasonic sound waves by subjecting crystals to an oscillating electric current. In 1917, during World War I, a pupil of Curie's, the French physicist Paul Langevin (1872-1946), worked out a method of sending out a beam of ultrasonic sound and detecting the echo. This is called "echo-location," or "sonar," the latter being an acronym for "sound navigation and ranging," where "ranging" means "determining distance."

The immediate purpose of Langevin's device was to detect enemy submarines, but by the time he had smoothed out the last difficulties in the way of such use, the war was over.

Sonar could, however, be used for more important tasks in peacetime. It could be used to measure the distance to the sea bottom in a matter of seconds, with no fear of the distorting influence of currents. Furthermore, it could give a continuous record of every bit of the bottom over which a ship traveled. More detail could be gathered in five minutes than the *Challenger* could manage in its entire voyage.

The first ship to use sonar in this way was the German oceanographic vessel *Meteor*, which studied the Atlantic Ocean in 1922. It soon became obvious that the ocean bottom was by no means featureless and flat and that Telegraph Plateau was not a gentle rise and fall. Telegraph Plateau was a mountain range, longer and more rugged than any mountain range on land. It wound down the length of the Atlantic, and its highest peaks broke through the water surface and appeared as islands, such as the Azores, Ascension, and Tristan da Cunha.

Later soundings elsewhere showed that the mountain range was not confined to the Atlantic Ocean. At its southern end it curves around Africa and moves up the western Indian Ocean to Arabia. In the middle of the Indian Ocean, it branches, so that the range continues south of Australia and New Zealand and then works northward in a vast circle all around the Pacific Ocean. What began in men's minds as the Mid-Atlantic Ridge quickly became the Mid-Oceanic Ridge.

After World War II, the details of the ocean floor were probed with new energy by two American geologists, William Maurice Ewing (1906–74) and Bruce C. Heezen (1924–77). Detailed soundings in 1953 showed a deep canyon running the length of the ridge and right along its center. This was the beginning of the notion that Earth's crust was split into a number of large "plates" that slowly moved relative to each other. This was the basis of "plate tectonics," which revolutionized geology, showing that mountain-building, volcanoes and earthquakes, and even the course of biological evolution depended on the slow motion of the plates.

The new knowledge of the oceans casts the known geography of the land into a truer light. For instance, Mt. Everest, which has a height of 8,847 meters above sea level, is the tallest mountain on Earth—if we measure from sea level. Mt. Everest, however, stands on the Tibetan Plateau, which is itself at an average height of 4,800 meters above sea level. If Mt. Everest is measured from its base rather than from a theoretical sea level, it is only about 4,000 meters high.

Consider, in contrast, the island of Hawaii, which is essentially a mountain that rises out of the water. Its highest peak, Mauna Kea, reaches to only 4,205 meters above sea level, but it stands on the Pacific sea bottom and most of it is hidden by the ocean. From its base to its peak, Hawaii has a height of 10,000 meters and is, on that basis, the tallest mountain in the world.

If we could imagine a dry Earth with its ocean basins empty, the rise of Hawaii from that dry bottom would be unimaginably impressive and it would be seen without difficulty to be the largest and tallest of all earthly mountains. (Nevertheless, even so, Hawaii wouldn't be as difficult to climb as Mt. Everest, since the latter, higher with respect to sea level, would have a peak that was much colder and immersed in much thinner air than the peak of Hawaii.)

There are also deep abysses in the ocean, vast "trenches," compared to which the Grand Canyon is a scratch in the ground. These abysses, with sea bottoms far deeper than the average depth of the ocean, are the oceanic extreme.

The trenches, all located alongside island chains, have a total area amounting to nearly 1 percent of the ocean bottom. This may not seem much, put that way, but it is actually equal to half the area of the United States.

There is a trench off the Indonesian islands in the Indian Ocean. In the Atlantic Ocean, there are trenches off the West Indies and the South Sandwich Islands. The deepest trenches, not surprisingly, are in the giant Pacific Ocean, where they are found off the Philippines, the Solomons, the Kuriles, the Aleutians, and the Marianas.

The average depth of the Pacific Ocean is about 4,250 meters, but the depth of the trenches is about twice that. In 1951, a British oceanographic vessel, *Challenger II,* found that the Marianas Trench was the deepest. It found the lowest part of it (now called the Challenger Deep) and measured that depth to be 10,900 meters. In 1959, a Soviet oceanographic vessel reported that another spot of the Challenger Deep was at a depth of 11,033 meters.

The deepest ocean deep, then, is considerably farther below sea level than the highest mountain peak is above it. If Mt. Everest, together with the base it stands on, were imagined fitted into Challenger Deep, 2,186 meters of ocean water would roll over its head, and prior to World War I its very existence would not have been expected unless some lucky sounding had been made right over it. Even the highest peak of Hawaii would not rise above the surface of the Pacific if it were standing in the Challenger Deep.

The total vertical distance between the deepest deep and the highest height, then, is 19,880 meters, and human beings have now been able to measure it all. If there are parts of either land or sea bottom that have not been plotted out in detail, it is only because there has not yet been time or occasion to do it. Once attention is drawn to a particular missing spot, it can be filled in without trouble.

## VESSELS OF THE DEEP

To measure a depth is one thing. To send human beings down into such depths is quite another, and a much more difficult task, too. Exploring the ocean deep is more difficult than climbing a mountain, for instance, if only because water is even less breathable than thin air is.

Scuba-diving apparatus suffices only for the region just below the surface. Diving suits, with airtight helmets into which air is pumped, are scarcely any better as far as depth is concerned and are much clumsier in maneuvering.

Diving bells or "caissons" can be used for working underwater. Essentially, they are structures closed on the top and open on the bottom. The water rising into them traps air. Devices for the renewal of the air can be used, but the air must be compressed till its pressure is equal to that of the surrounding water.

As mentioned earlier, nitrogen under pressure dissolves in the body fluids to a greater extent than under normal conditions. If the pressure is then reduced rapidly—that is, if someone who has been down deep is simply hauled up—the nitrogen becomes less soluble with decreasing pressure and bubbles out of solution, collecting in joints and in small blood vessels, inducing agonizing pain and sometimes, death. Such "bends" or "decompression sickness" can only be treated by recompression to dissolve the nitrogen again and then a very slow decompression to allow the nitrogen to bubble out very slowly and be removed by the lungs in stages. It is better to prevent the ailment altogether by very slow decompression in the first place.

The risk of bends is somewhat lowered if an artificial gas mixture of helium and oxygen is used, since helium dissolves in the body fluids to a considerably lesser extent than does nitrogen. Still, the use of compressed air increases the hazards of underwater work and, of course, limits the depth to which one can penetrate.

What is needed is a solid-walled ship, capable of withstanding water pressure, which can contain air at normal pressure. If the ship is maneuverable as well, it is a "submarine."

The concept is a natural one, and a number of people had it in early modern times. A Dutch inventor, Cornelis Drebbel (1572–1634), is the first who is supposed actually to have built one. He constructed it of greased leather over a wooden frame, and between 1620 and 1624 he maneuvered it four or five meters under the surface of the Thames River.

The most forceful drive behind repeated attempts to build submarines was its clear usefulness in war. Underwater, such a vessel would be invisible and could approach an enemy ship unseen and, for instance, attach explosives to it below the waterline.

Thus, an American inventor, David Bushnell (1742–1824), built a one-man submarine during the American Revolutionary War and with it tried to sink a British warship in New York Harbor. The submarine was of wood and was moved by a propeller cranked by hand from inside. The device worked well enough, but Bushnell could not drill through the copper sheathing around the hull of his quarry. He tried again during the War of 1812 in the harbor of New London, Connecticut, and again failed.

Robert Fulton (1765–1815), who was eventually to gain fame as the constructor, in 1807, of the first steamship sufficiently successful to establish the predominance of steam power on the water, earlier worked on a submarine, which he named the *Nautilus*.

He was the first to make use of metal for the purpose, building his vessel of copper sheets over iron ribs. He made it large enough to give it an air supply sufficient to support four men and two candles for three hours. There was a glass-covered porthole for observation and, as in Bushnell's case, a hand-cranked propeller for propulsion.

He built it in 1800, when France and Great Britain were at war, and Napoleon Bonaparte, the French dictator, was interested for a while. Unfortunately, while Fulton's submarine might well have sunk any British ship, it didn't move fast enough to catch any, and Bonaparte lost interest. Nor could Fulton interest the British, whom he next dealt with.

Fulton came back to the United States and planned to build a steam-powered submarine large enough to carry a hundred men, but died before the craft could be finished.

During the American Civil War, the Confederacy, faced with a strangling blockade by the Union navy, sought a variety of ways to break the blockade. One of them was to build submarines, a project headed by Horace Lawson Hunley (1823–63) of Alabama. Three submarines were built and suffered a series of failures, in one of which Hunley himself died.

The third submarine, named *Hunley,* continued in use after Hunley's death. On February 17, 1864, it finally got the first kill ever to be attributed to a submarine. It managed to sink the Union warship *Housatonic* in the harbor of Charleston, South Carolina, with a loss of five men. The explosion, however, also destroyed the submarine and its crew.

The French science fiction writer Jules Verne (1828–1905) built the most famous submarine of all time—though in literary fashion only—in his novel *Twenty Thousand Leagues Under the Sea,* published in 1870. He had followed the course of the American Civil War, and it is quite likely that the Confederate attempts to build submarines inspired him to produce his own famous *Nautilus* (the same name Fulton had used for his submarine seven decades before).

A submarine couldn't become really practical till some way could be found to move it by other than a hand-cranked propeller. A steam engine couldn't really be used underwater because it would have to be run by burning fuel, and this would quickly exhaust the submarine's air. What was needed was a motor run by electricity from a storage battery.

The first electric submarine was again named *Nautilus* and was built in 1886 by two Englishmen. The battery had to be periodically recharged, of course, and the vessel's cruising distance between recharges was something like 130 kilometers.

Many submarines were then built, using steam power and, later, diesel engines when cruising at the surface and electricity when submerged. The American engineer Simon Lake (1866–1945) built the submarine *Argonaut* in 1897. It sailed from Norfolk, Virginia, to New York under its own power and was the first submarine to operate successfully in the open sea.

By the time World War I began, the major European powers all had submarines. From 1914 to 1945, submarines were continually improved and were used almost exclusively for war or readiness for war. Yet throughout this period they remained essentially surface vessels. They could submerge for periods of time, but invariably had to surface frequently to recharge their batteries.

A fundamental advance came on January 17, 1955, when the first nuclear-powered submarine was launched. It was (once again) the *Nautilus,* and its great advantage over nonnuclear submarines was that its power source would not be exhausted for long periods of time so that it could remain submerged for months, if necessary. The Russians built their first nuclear submarine in 1960, the British in 1963, and the French in 1969. The nuclear reactors on such submarines could operate for 640,000 kilometers (sixteen times Earth's circumference) before having to be overhauled.

Nonnuclear submarines can reach depths as great as 2,500 meters, well below the depth any whale can reach, and nuclear submarines can possibly penetrate to depths of over 6,000 meters.

The ability of submarines to penetrate the depths is merely incidental, since their primary function is war-making. But even before submarines were outfitted with nuclear reactors, devices were built that were intended only for the study of the ocean depths. The pioneer in this respect was an American naturalist, Charles William Beebe (1877–1962), who was interested in corals and wanted to pursue them to their haunts deep under the ocean surface.

He designed a watertight sphere of thick metal with a window of thick quartz that would be capable of withstanding the enormous pressures of deep water. It was spherical in shape for maximum strength (at the suggestion of Governor Franklin Delano Roosevelt of New York, a friend of Beebe's). It carried compressed oxygen and chemicals to absorb carbon dioxide.

Beebe called it a "bathysphere," from Greek words meaning

"sphere of the deep." The bathysphere was not maneuverable. It was merely a metal-enclosed bubble of air, barely large enough to hold two men. It dangled from a surface ship by a metal cable. If storms, winds, and waves battered the ship, or if anything else happened that might snap the cable, the bathysphere would drop to the ocean bottom and anyone inside could not possibly survive.

Beebe and a companion, Otis Barton, nevertheless took a chance. In their first dive, on June 11, 1930, they reached a depth of 400 meters. In 1932, they reached a depth of 920 meters. Beebe made over thirty dives altogether and then abandoned the task as unproductive of any further important information.

Barton, in a modified bathysphere, plumbed to a depth of 1,370 meters in 1948.

What was needed, clearly, was a maneuverable bathysphere. To begin with, it would have to float despite the heavy metal required to hold off water pressure. A Swiss physicist, Auguste Piccard (1884–1962), who had met Beebe in 1933, got the notion of combining the bathysphere with the principle of the balloon. Balloons were suspended by a volume of lighter-than-air gas. What was needed was a bathysphere suspended by a volume of lighter-than-water liquid. Piccard used gasoline for the purpose, placing it in a ship-shaped float large enough to dwarf the spherical bathysphere affixed firmly to its bottom.

The result was a "bathyscaphe." ("ship of the deep"). The cigar-shaped float contains thirteen tanks, eleven of which are filled with gasoline and two of which are empty. The float can just lift the weight of the bathysphere and keep it from sinking.

The two empty tanks in the float can be opened and allowed to fill with seawater. The extra weight of the seawater drags down the bathyscaphe and causes it to sink downward to the bottom of the sea.

If it is sinking too quickly, there are up to 13 tons of small iron pellets attached to the sphere that can be gradually released. This makes the bathyscaphe lighter so that it sinks more slowly. If enough pellets are released, the bathyscaphe rises back to the surface.

Once the bathyscaphe has gone down and then returned to the surface, the seawater can be pumped out of the tanks and a new supply of iron pellets can be taken on to make it ready for another descent and ascent.

Piccard had to wait till after World War II before he could actually build his bathyscaphe. In 1948, the first one was completed. It was tested, rebuilt, and improved, and, on February 15, 1954, in the first real test dive off the coast of West Africa, two French naval officers descended to a depth of 4,050 meters and returned safely.

Then a still better bathyscaphe was built, the *Trieste*, and in 1958 it was bought by the United States Navy. It was taken to California and was still further improved. In due course, its successor, the *Trieste II*, was ready for the big test.

Out it went to the western Pacific. Ready to board it were Jacques Piccard (1922–      ), the son of Auguste, and an American naval officer, Don Walsh.

At 8:20 A.M. on January 23, 1960, the *Trieste II* was positioned over the Marianas Trench and sank downward for 10,915 meters to the bottom, coming to rest on soft mud. The mud billowed up and obscured the view for a while, but it slowly settled, and as visibility got better, what the two men saw in their searchlights was a small red shrimp, about an inch long, floating by. They also saw a one-foot-long flounderlike fish.

The bathyscaphe then jettisoned its iron pellets and rose to the surface. The two men were safely back at 5:00 P.M. after a very dangerous 22-kilometer journey that took them nine hours.

The exploration of the deep sea has presented scientists with some exciting surprises.

As it happens, at the juncture of the tectonic plates there are occasional "hotspots" through which heat leaks into the ocean, rather like hot springs on dry land. Such hotspots were first suspected in 1965 and were detected in the early 1970s.

Beginning in 1977, a deep-sea submarine carried scientists downward to investigate such hotspots. At the mouth of the Gulf of California, they found "chimneys" through which hot gushes of smoky mud surged upward, filling the surrounding seawater with minerals.

The minerals are rich in sulfur, and the neighborhood of these hotspots is full of special kinds of bacteria that obtain their energy from chemical reactions involving sulfur plus heat, instead of from sunlight. Small animals feed on these bacteria, while larger animals feed on the smaller ones.

This was the discovery of a whole new chain of life forms that do not depend on sunlight and plant cells, but they can exist, of course, only near the hotspots. The scientists found clams, crabs, and various kinds of worms, some of them quite large. They are special species that can live in water so filled with chemicals that it would be poisonous to other forms of life.

And so the human horizon expanded to include the entire ocean to its uttermost depth, as earlier it had expanded to include the entire surface, both land and sea.

# Interlude:
## The Horizon
## of Numbers

## LARGE NUMBERS

In this book, in which I investigate horizons of one sort or another that have been penetrated and expanded by the human mind, it is inevitable that I will eventually use very large and very small numbers. So far I have avoided doing so, but a few are about to come up, and before that happens, it seems only reasonable to consider how humanity has pushed back the horizons of numbers themselves.

In prehistoric times, human beings did not have need for large numbers. There are primitive tribes, even today, with very few number words. One hears of some tribes which have names equivalent to only "one" and "two." For any number over two, they will speak of "many."

This doesn't mean, of course, that members of such tribes are incapable of recognizing three objects or four. They just don't have special names for such numbers, but they can get along with what they have. If a member of such a tribe lends a friend "two and two more" of an object, he will know the difference and be seriously annoyed if he gets back "two and one more."

As a matter of fact, in ancient and medieval times, our ancestors did not do very much better. We had only twelve different number names, and all the many numbers that were used were built up out of those twelve. Ten of the different number names are, naturally, for the first ten numbers: one, two, three, four, five, six, seven, eight, nine, ten. (These are the names in English. Every language has its own names.) Presumably, we have ten different names because we have ten fingers on our two hands and if we match numbers against fingers, it is only natural to give each finger a different name.

Beyond that we have derived names. "Eleven" is a distortion of an Anglo-Saxon expression that corresponds to "one left" in modern English—one left, that is, after ten have been counted off, making a total of eleven. In the same way, "twelve" is "two left." "Thirteen" is a distorted form of "three and ten," "twenty" is a distorted form of "two tens," and with this in mind we can see the meaning of any number in terms of the basic ten, up to "ninety-nine" ("nine tens and nine").

Following this system, the number after "ninety-nine" would be "tenty" and after "tenty-nine" we would have "eleventy," and so on. Children, recognizing the logic behind this, sometimes make use of such numbers and are quickly corrected by their elders who know that the number after "ninety-nine" is "one hundred." The name "hundred" has its origins back in prehistory, as does "thousand," which comes after "nine hundred ninety-nine." "Hundred" and "thousand" are the two remaining number names that are of early origin.

The Roman word for "thousand" was *mille,* from which is derived our "mile," which was, originally, 1,000 Roman paces in length.

People can use numbers well beyond "one thousand," of course, without having to make up new names. We can speak of "ten thousand" or "nine hundred thousand" or "a thousand thousand," for that matter.

The Greeks had a word *myrios,* meaning "innumerable" and from this they formed the word *myrias* to mean "ten thousand." We get our word "myriad" from the Greek term and use it to mean "a very great number," though on rare occasions someone may use it to mean precisely "ten thousand."

The Greek mathematician Archimedes (287–212 B.C.) wrote a treatise intending to show that the number system could be used to enumerate any quantity, however large, and to do so he used the scheme of talking of myriads of myriads of myriads—he did not try to simplify matters by inventing a name for a number larger than a myriad.

In the late Middle Ages in Italy, commerce and wealth grew to the point where, in bookkeeping, it became necessary with considerable frequency to speak of so many "thousand thousands." The Italian merchants found it useful to invent a special word for "thousand thousand." They called it *milione,* meaning "a large *mille*"—that is, "a large thousand." The name is "million" in English. (It was about this time, too, that Arabic numerals were coming into widespread use and the symbols 1, 2, 3, 4, 5, 6, 7, 8, and 9 were adopted.)

The use of "million" became general in Europe by the 1500s, and by that time a still higher number word had come into use. The French mathematician Nicolas Chuquet (1450–1500) was the first to use "billion" for "million million." The prefix "bi" is from the Latin *bis,* meaning "twice." "Billion" is a number, in other words, in which "million" appears twice.

In Arabic numerals, "one hundred" is 100, "one thousand" is 1,000, and "one million" is 1,000,000. If "one billion" is "one million million," it is written 1,000,000,000,000. That is the meaning of "one billion" in Great Britain and Germany today. In the United States and France, however, "one billion" has come to be used for "one thousand million" or 1,000,000,000.

Once "billion" was well established, still other words were coined, making use of Latin prefixes for three, four, five, six, and so on. We therefore have words such as "trillion," "quadrillion," "quintillion," "sextillion," "septillion," "octillion," etc.

In American usage, each of these names is a thousand times the value of the one before, so that a trillion is a thousand billion (1,000,000,000,000), a quadrillion is a thousand trillion (1,000,000,-000,000,000), and so on. In British usage, each of these names is a million times the former, so that a trillion is a million billion or a million million million (1,000,000,000,000,000,000), while a quadrillion is a million trillion (1,000,000,000,000,000,000,000,000).

The largest number name that is ordinarily to be met with using this system is "centillion." The prefix is from the Latin *centum,* meaning "hundred," so that a centillion in the American system has a hundred groups of three zeroes in addition to the basic three for "thousand" and is therefore a 1 followed by 303 zeroes. In the British system it would be a 1 followed by 600 zeroes.

It would, of course, be easy to invent new number names to any indefinite extent, but it would be pointless to do so, for scientists have devised a much more sensible and convenient system for referring to large numbers.

The French mathematician René Descartes (1596–1650) grew tired of representing repeated multiplications of a particular number or symbol such as $a \times a$, or $a \times a \times a$. He preferred to represent such products as $a^2$ and $a^3$. The figure above the line is called an "exponent," and such "exponential notation" came into general use.

Since $10 \times 10 = 100$, and $10 \times 10 \times 10 = 1,000$, it is possible to refer to 100 as $10^2$ and to 1,000 as $10^3$. It is easy to see that 10,000 is

$10^4$, 100,000 is $10^5$, 1,000,000 is $10^6$, and so on. What makes this system easy to remember and use is that the exponent represents the number of zeroes after the 1 in such numbers. Thus we know at once that 10,000,000,000,000 is $10^{13}$ and that $10^{17}$ is 100,000,000,000,000,-000. Again, once we define a centillion as a 1 followed by 303 zeroes, in the American system, we also know it is $10^{303}$. In the British system, it is $10^{600}$.

While there is a chance for confusion in using the number names, since all the number names higher than a million signify different values in different nations, exponential notation is generally agreed on everywhere and there is no ambiguity.

Exponential notation can be used for numbers generally. Thus, 6,300 can be written as $63 \times 100$, or $6.3 \times 1,000$, or even $0.63 \times 10,000$. It can therefore be written as $63 \times 10^2$, or $6.3 \times 10^3$, or $0.63 \times 10^4$. Each of these alternatives is correct, but it is customary to choose the one in which the non-exponential portion of the figure is between 1 and 10. Therefore 6,300 should be written $6.3 \times 10^3$.

In writing or thinking of large numbers, we can easily see that there is no way of writing or thinking of such a thing as "the largest number." Any number you can have the patience to write or to think of can always have 1 added to it, or it can always be doubled, or it can always have a group of zeroes added to it.

The early Greek mathematicians must have understood this, but it was not until late-medieval times that mathematicians began to deal with unending series of numbers that grew steadily larger and never stopped growing larger.

In 1202, for instance, an Italian mathematician, Leonardo Fibonacci (1170–after 1240), first introduced what is now called the "Fibonacci series." This is a sequence of numbers beginning with 1, 1, where each number is the sum of the two preceding. The sequence, then, is: 1, 1, 2, 3, 5, 8, 13, 21, 34, 55, 89, and so on forever.

This is an unending series of numbers, or an "infinite series," where "infinite" is from a Latin word meaning "unending."

By the 1680s, when the English mathematician Isaac Newton (1642–1727) introduced a branch of mathematics called "the calculus," infinite series played an exceedingly important role and the question of how to symbolize them became important. In the case of the Fibonacci sequence, I gave the rule for generating successive numbers, then listed the first eleven numbers and said "and so on forever."

Mathematicians, writing hurriedly, needed something more compact. For instance, the simplest sequence is that of natural numbers: 1, 2, 3, and so on forever. You might write it: 1, 2, 3, . . . In that way, you let the series of dots indicate you are continuing the list forever.

That is perhaps not sufficiently specific. You might write 1, 2, 3, . . . , 20 meaning you are including all the numbers from 1 to 20 without omitting any, but without taking the trouble to write them all down. You could also have 1, 2, 3, . . . , 200, or you could have 1, 2, 3, . . . , 10,000,000. But how can you express in this way *all* the numbers without end?

In 1656, an English mathematician, John Wallis (1616–1703), used a sideways 8 as the symbol. If you write 1, 2, 3, . . . , $\infty$, you mean the sequence is viewed as being continued forever. The symbol $\infty$ means "and so on forever" or "and so on endlessly" or "and so on infinitely."

The symbol $\infty$ is often read as "infinity," meaning "endlessness," and there is a strong temptation to consider it a number—specifically, the largest number there is. That, however, is wrong. The symbol $\infty$ is not a number but a characterization of the property of endlessness.

Actually, there are different degrees of endlessness. Thus, you can count anything (you might think) by the use of the sequence of natural numbers. All the grains of sand on all the beaches of the world, all the stars in the sky, all the atoms in all the universe, and all the seconds that the universe has existed can be enumerated without exhausting the number system and without getting any closer to "infinity" than you were when you started.

Yet not so! A German mathematician, Georg Cantor (1845–1918), was able to show, in 1874, that there was no way of counting all the points in a line, for instance. No matter what system was used to count them, an infinite number of them were always omitted. The entire endless series 1, 2, 3, . . . , $\infty$ did not suffice to count them.

This meant that the infinite number of points in a line represented a higher order of infinity than the infinite number of numbers in the 1, 2, 3, . . . , $\infty$ sequence. In fact, Cantor showed that it was logical to suppose that there was an endless series of orders of infinity, of which the endlessness of numbers was the very least.

The imagination staggers, but in this book, it may be a relief to realize that we will deal only with finite numbers and need go no further into the matter of endlessness.

## SMALL NUMBERS

Let us move off in the other direction. How *small* can numbers become?

If we deal only with the natural whole numbers, 1, 2, 3, . . . , ∞, then 1 is the smallest number. You can't very well have fewer pebbles than one, for instance.

Except that you can. You create smaller and smaller numbers by subtraction. If you begin with five pebbles and take 1 away, you have four. If you repeat the process, you have three, then two, then one. But you can repeat the process once more, take away that one last pebble, and have none at all.

It might be argued that "nothing" is not a number, but mathematicians have found that if nothing is given a symbol, that symbol can be manipulated like any other number symbol (except that it is impossible to divide by "nothing," so that process is forbidden).

Throughout ancient times, no one, as far as we know, ever thought of giving "nothing" a symbol. Without that symbol, no really useful system of number symbols could be evolved. The Egyptians, Babylonians, Hebrews, Greeks, and Romans all used symbols that required clumsy manipulation and that made progress in mathematics almost impossible.

Many scientists used the abacus to manipulate numbers, and to represent "one hundred two" they would have one counter in the hundreds column and two in the units column, but none in the tens column. They never thought of using a symbol to represent "none in the column."

The idea occurred to a mathematician in India (one forever nameless and unknown) sometime before A.D. 876. The concept of "nothing" was represented by a little dot, as being the closest still-visible approach to nothing; or as a little circle, which is just a hollow dot and therefore an even closer approach to nothing. We write it 0, so that one hundred two is written 102, with the 0 indicating "the absence of counters in the tens column."

The Arabs borrowed the notion from the Indians and passed it on to Europe. The Arabs called the symbol *sifr*, meaning "empty" (the empty column in the abacus, you see). From this we get "cipher" for working with numbers, or for something or someone that is of no account, or for a hidden message (that must be ciphered out). The cipher makes it possible to distinguish between 12, 102, and 120, for instance, and made the Arabic numeral system with its positional notation far superior to anything that had gone before.

From *sifr* also comes "zero," our own name for the symbol.

It might seem that zero is now the smallest number without question, since what can be smaller than nothing?

But consider. Suppose you had no money and, in addition, owed someone a dollar. You would then have less than nothing, for if somehow you earned a dollar you would have to give it away to cancel your debt. And, of course, you could have a debt larger than a dollar and therefore have less than nothing to an even greater degree.

We have "negative numbers" representing such debts, and we can write these $-1$, $-2$, $-3$, . . . , $\infty$, where these numbers are read "minus one," "minus two," "minus three," and so on. The first to recognize the general mathematical usefulness of negative numbers was the Italian mathematician Girolamo Cardano (1501–76).

Yet though negative numbers are less than nothing, they are not necessarily small numbers. If you owe a million dollars, you would have far, far less than nothing; but it would not appear to you as though you had very little money. It would appear to you as though you had a very large debt.

It is more common to look upon small numbers as numbers that are very close to zero, whether on the ordinary side or on the negative side. You may have very little money (though not none at all) or a very small debt (though not none at all).

This brings us to the concept of fractions. You can have one of an object, but if that object can be divided, then you can have half of an object, or a quarter of it, and so on. If you divide an object into two equal pieces, then each piece is the result of one divided by two or, in modern symbols, $1/2$, where the "slash" (/) represents "divided by."

Similarly, you can have $1/3$ or $1/4$ or $1/5$, and so on. The greater the number of equal pieces into which you divide the original unit, the smaller the individual piece. Mathematically, you can divide a unit number into any number of pieces so that you end up with the sequence $1$, $1/2$, $1/3$, $1/4$, . . . , $1/\infty$.

Just as there is no such thing as a largest number in mathematics, so there is no such thing as a smallest number. The larger the number in the denominator (the figure to the right of the slash), the smaller the number, and since you can always make the denominator larger, no matter how large it was to begin with, you can always make the number smaller, no matter how small it was to begin with.

You can express very small numbers exponentially, just as you can express very large numbers so.

Thus $1/100$ can be written $1/10^2$, and $1/1,000,000$ (a millionth) can be written $1/10^6$, and so on.

It would be convenient if we could avoid the fractional form, and, oddly enough, we can.

Thus, 10,000 divided by 10 equals 1,000, which divided by 10 equals 100, which divided by 10 equals 10. In other words, $10^4$ divided by 10 equals $10^3$, which divided by 10 equals $10^2$, which divided by 10, equals $10^1$. Each time you divide an exponential ten by ten, you decrease the size of the exponent by one.

Therefore $10^1$ divided by ten should be equal to $10^0$. We know that $10^1$ is equal to 10, and 10 divided by ten is equal to 1. It follows, then, that to prevent contradiction, $10^0$ is equal to 1.

Again, $10^0$ divided by ten is equal to $10^{-1}$, which divided by ten is equal to $10^{-2}$, which divided by ten is equal to $10^{-3}$, and so on forever. Since 1 divided by ten is $1/10$, which divided by ten is $1/100$, which divided by ten is $1/1,000$, and so on, it follows that $10^{-1} = 1/10$; $10^{-2} = 1/100$; $10^{-3} = 1/1,000$, and so on forever.

We can also see that if a million is $10^6$, then one-millionth is $10^{-6}$, and if a trillion is $10^{12}$, then one-trillionth is $10^{-12}$, and so on.

Suppose you express the number in decimal form. With ordinary exponents, the exponent (as I have said before) is equal to the number of zeroes after the 1. With negative exponents, the exponent is equal to the number of zeroes before the 1, *provided you include one zero before the decimal point.*

Thus, $10^1 = 10$, while $10^{-1} = 0.1$, and $10^8 = 100,000,000$, while $10^{-8} = 0.00000001$, and so on.

What if you have a number like 0.0000638? That is equal to 6.38 × 0.00001, and it can be written as $6.38 \times 10^{-5}$.

Now we have enough of an armory of numbers to go on to the ever-receding horizons of human exploration and thought.

# 6
# Below Earth's Atmosphere

## THE SOLID INTERIOR

The surface of Earth, whose horizons we have been discussing thus far, has a 20-kilometer range of unevenness from the height of Mt. Everest to the depth of Challenger Deep, but this range shrinks to insignificance in comparison with the size of Earth itself.

Earth is a ball that is over 12,000 kilometers in diameter, and if it were modeled into an object the size of a billiard ball, with all its surface unevennesses reproduced exactly to scale, the model would be *smoother* than an ordinary billiard ball—and the ocean would be an all but unnoticeable mist of dampness over 70 percent of its surface.

The expansion of the human horizon through mountains and oceans is *still* essentially two-dimensional, therefore. How far, then, can one extend human knowledge beyond the surface, really beyond? Do we know anything, for instance, about the interior of the solid sphere of Earth itself?

Yes, for we know something about Earth as a whole just from our knowledge of the surface.

From the manner in which ships at sea disappear over the horizon, from the way stars appear and disappear over the horizon as one travels north and south, and from the shape of Earth's shadow on the moon, people have realized since the time of the ancient Greeks that Earth is essentially spherical in shape. Since Earth rotates, there has to be a centrifugal effect, as Isaac Newton (1642–1727) was the first to point out in the 1680s, and the surface must bulge outward more and more as one moves away from the poles.

the bulge reaching a maximum outward extension at the equator. The result is that Earth is an "oblate spheroid" rather than a sphere. However, the bulge is so small compared to the size of Earth itself that seen from space, Earth would seem to be a sphere.

Thus, the diameter of Earth is 12,714 kilometers from pole to pole, and 12,757 kilometers across the equator. The equatorial bulge, at its maximum, adds 43 kilometers to the diameter, a difference of 0.32 percent.

The volume of such an oblate spheroid is $1.083 \times 10^{12}$ cubic kilometers,* or, in other words, about a trillion cubic kilometers.

Somewhat more difficult to determine is the "mass" of Earth, where we might define the mass of an object (not quite accurately, but sufficiently so for our purposes) as the quantity of matter in that object.

One way of judging the mass of an object is to measure the intensity of its gravitational field, since that is strictly proportional to its mass. In 1798, the English scientist Henry Cavendish (1731–1810) very delicately measured the tiny gravitational pull of one lead sphere of known mass upon another and compared this with Earth's pull on those masses. From the much greater pull of Earth, he could calculate Earth's much greater mass.

Cavendish's initial figure for the mass of Earth was surprisingly accurate, and it has been little improved on since. The figure accepted today is $5.976 \times 10^{24}$ kilograms † (which is just about 6 trillion trillion kilograms).

If we divide Earth's mass by its volume, we obtain Earth's average "density"—the mass of a unit volume of Earth. The unit volume we will use is the cubic meter.‡ The average density of Earth comes out to 5,518 kilograms per cubic meter, or (in more compact symbols) 5,518 kg/m³.

It is common to compare the density of an object with that of water, which has a density of just 1,000 kg/m³. The average density of Earth, then, is just 5.518 times that of water.

The surface rocks of Earth have an average density of about 2,800 kg/m³. Therefore the deeper regions of Earth must have a density considerably higher than 5,518 kg/m³ in order to produce that as the average.

---

* A cubic kilometer equals 0.24 cubic miles in common American measure, so that there are just about 4 cubic kilometers to a cubic mile.

† A kilogram is equal to just about 2.2 pounds in common American measure, so that 5 kilograms equals 11 pounds.

‡ A cubic meter is equal to about 1.3 cubic yards in common American measure, so that 3 cubic meters is equal to about 4 cubic yards. Also, a cubic meter is equal to one-billionth ($10^{-9}$) of a cubic kilometer.

If Earth were uniformly rocky throughout, we would expect the density to rise with depth as the weight of thicker and thicker layers of rock above squeezed the rock below into greater and greater densities.

The pressures that would build up even at Earth's center would not be high enough, however, judging from data gathered from laboratory experiments, to compress rock into densities sufficiently high to produce the calculated average. The conclusion, then, is that Earth is not uniformly rocky throughout; somewhere near the center there is a change in structure, some substance that is denser than rock must be involved.

As it happens, meteorites are occasionally found on Earth. These represent small pieces of planetary material that ordinarily circle the sun and have happened to collide with Earth in the course of that circling. They come in two chief varieties. Some are rocky and are in many ways similar to the kind of rock with which we are familiar here on Earth. Some, on the other hand, are metallic, and are made up of iron and the similar metal, nickel, in a roughly 9-to-1 ratio.

One theory about meteorites (not the one most favored today) held that they had their origin in the disintegration of a planet. If so, perhaps these two varieties of meteorites represent the two varieties of solids that make up planets. In 1866, a French geologist, Gabriel Auguste Daubrée (1814-96), suggested that the interiormost portion of Earth was made up of an iron-nickel mixture like that of the metal meteorites.

This seemed logical, and it has remained the favored suggestion even today. If so, the density at Earth's center should be about 11,500 kg/m³, or just about twice that of the surface rocks.

## EARTHQUAKES

But what about the details concerning the change in density as one moves downward? If the structure of Earth changes, if there is a more or less sudden switch from rock to metal, there should also be a sudden change in density. How can one check the solid structure of Earth for the location and extent of this sudden change? To do so, one must find something that penetrates the deep layers, that can be detected after having done so, and that can experience changes in doing so that will yield the desired information.

As it happens, there *is* such a phenomenon.

The slowly moving plates of Earth's crust may bind and then

suddenly slip as they rub against each other. These slips set up powerful vibrations we call earthquakes. The strength of the vibrations falls off with distance, and so actual damage is done only in the immediate vicinity of the slippage. The vibration can, however, be detected a great distance—virtually everywhere on Earth, if the earthquake is sufficiently violent.

The earth suffers a million quakes a year, including at least ten disastrous ones and a hundred serious ones, so this is a phenomenon that forces itself on the attention of scientists.

The English geologist John Michell (1724–93) was the first to suggest, in 1760, that earthquakes were waves set up by the shifting of masses of rock kilometers below the surface, and in 1855 the Italian physicist Luigi Palmieri (1807–96) devised the first "seismograph" to study these waves.

In its simplest form, the seismograph consists of a massive block suspended by a comparatively weak spring from a support firmly fixed in bedrock. When the earth moves, the suspended block remains still because of its inertia. However, the support moves with Earth's motion. This motion is recorded on a slowly rotating drum by means of a pen attached to the stationary block. Nowadays, seismographs are apt to use a ray of light marking light-sensitive paper, in order to avoid the drag of pen on paper.

The English engineer John Milne (1850–1913), using seismographs of his own design, showed conclusively in the 1890s that Michell's description of earthquakes as waves propagated through the body of the earth was correct. He led the way to establishing seismograph stations here and there on Earth. Well over 500 are now spread over every continent, including Antarctica.

Seismograph studies showed that earthquake waves come in two general varieties, "surface waves" and "bodily waves." The surface waves follow the curve of Earth's surface, while the bodily waves go through the interior. The path through the interior is, so to speak, a shortcut, and so bodily waves arrive first at the seismograph. The extent to which the bodily waves beat the surface waves to a particular seismograph is an indication of how far away the source of the wave (the "epicenter" of the earthquake) is.

The bodily waves are, in turn, of two types: primary ("P waves") and secondary ("S waves"). The primary waves, like sound waves, travel by alternate compression and expansion of the medium, rather like the pushing together and pulling apart of an accordion. Such waves can pass through any medium—solid, liquid, or gas.

The secondary waves, on the other hand, have the familiar form of snakelike wiggles at right angles to the direction of travel. They can travel through solids and along the surface of liquids, but they cannot travel through the body of a liquid or through a gas.

The primary waves move faster than secondary waves, and this too gives a hint as to the distance of the epicenter of an earthquake.

The speed of both the primary and secondary waves is affected by the properties of the material through which they pass. If the material is uniform in properties, the waves travel in straight lines. If the properties change, the waves curve in their path, and from the nature and extent of the curve, and from changes in the speed of travel, deductions can be made about the change in the properties of the material passed through. Thus, earthquake waves can be used as probes to investigate conditions deep under the earth's surface.

A primary wave near the surface travels at 8 kilometers per second (or, more concisely, 8 km/sec). At 1,600 kilometers below the surface, its velocity, judging from the arrival times, must be nearly 13 km/sec. Similarly, a secondary wave has a velocity of about 4.5 km/sec near the surface and 6.5 km/sec at a depth of 1,600 kilometers. Since increase in velocity is a measure of increase in density, we can estimate the density of the material far beneath the surface.

Whereas at the surface, Earth's rocky layer has an average density of 2,800 kg/m$^3$, 1,600 kilometers down it amounts to 5,000 kg/m$^3$, and 2,900 kilometers down to nearly 6,000 kg/m$^3$.

At a depth of 2,900 kilometers, there is an abrupt change. Secondary waves are not transmitted at all at lower depths. This is taken to mean that Earth's substance is liquid at lower depths. The change is a sharp one and there seems an abrupt alteration in properties—so abrupt that the boundary is called a "discontinuity." It is, in fact, the "Gutenberg discontinuity," named for the German geologist Beno Gutenberg (1889–1960), who in 1914 demonstrated its existence.

Below the Gutenberg discontinuity is Earth's "liquid core." At the greatest depths, within 1,300 kilometers of the planetary center, the core may be solid. This was first pointed out, from the behavior of earthquake waves, by the Danish geologist Inge Lehmann in 1936.

Above the Gutenberg discontinuity is Earth's "mantle."

The density of Earth leaps from about 6,000 kg/m$^3$ above the Gutenberg discontinuity to about 9,000 kg/m$^3$ below, if we go by earthquake-wave data, and this would indicate a sharp change in the chemical structure. This would fit with the notion that at the Gutenberg discontinuity, the rocky mantle suddenly changes into a liquid nickel-iron core. (There are currently arguments as to whether the core is entirely metallic or whether there is a certain admixture of sulfur or oxygen, and how much.)

Though the mantle is undeniably rocky in nature, it seems to differ from the surface rocks of Earth (if we judge by comparing the behavior of earthquake waves passing through both) in being richer in magnesium and iron, and poorer in aluminum.

The mantle, then, does not quite extend to the surface of Earth. A Croatian geologist, Andrija Mohorovičić (1857–1936), studying the waves produced by a Balkan earthquake in 1909, decided that there must be a sharp increase in wave velocity at a point about 32 kilometers beneath Earth's surface. This is now known as the "Mohorovičić discontinuity," and it marks the upper surface of Earth's mantle. Above it is Earth's "crust." (Core, mantle, and crust bear, coincidentally, just about the same relative proportions in volume as do the yolk, white, and shell of a hen's egg.)

Earth's crust is not evenly thick over the entire surface of Earth. The Mohorovičić discontinuity is farther beneath the surface in the continental areas than under the ocean. And on the continents, it sinks particularly low in the mountainous areas. There is, in fact, a rough symmetry, for the higher the land surface above sea level, the lower the Mohorovičić discontinuity. Thus the crust is up to 65 kilometers thick under mountain ranges, 35 kilometers thick in low-lying continental areas, and only 13 to 16 kilometers thick under oceanic areas.

The thickness of the crust in oceanic areas includes the thickness of the water overlying the seabed. Under the ocean deeps, then, the solid portion of the crust might be no more than 5 kilometers thick.

In the 1960s, there was some enthusiasm concerning the possibility of drilling through the seabed to the mantle in order to bring up material that would check the conclusions of theory. If that had taken place it would represent the deepest penetration into Earth's depths by human beings. The project fell through, however.

Nevertheless, even if the project is carried through some time in the future, it doesn't seem very likely that human beings will ever penetrate by instrument (much less by some sort of vehicle) lower than the boundary of the mantle. Nor can one honestly imagine what technological breakthrough would suffice for the purpose, considering the extreme properties of Earth's solid depths.

# 7
# Earth's Atmosphere

## ABOVE EARTH'S SURFACE

If the path downward into the bowels of Earth's solid structures seems to be one representing sharp limitations to human penetration, what about the path upward into the atmosphere?

In a way, human beings can and have explored atmospheric heights by climbing mountains. This, however, is an exploration that can take place only at the precise points on Earth's surface at which the mountains in question exist; and, in doing so, the human foot remains firmly planted on solid ground.

Is there no way in which human beings can declare their independence of the ground and rise bodily into the atmosphere? To do so has been a freedom-dream of human beings from earliest times, and the possible capacity to do so was suggested by what they could see about them.

The atmosphere *can* support objects. The wind can carry light objects through the air—leaves, feathers, seeds. More impressive is the fact that there are animals which are adapted to gliding long distances through the air. Examples are flying squirrels, flying phalangers—even flying fish.

Still more impressive are the true fliers, animals which by muscular effort can move through the air independently of air currents or even against them.

Four separate groups of animals have independently evolved mechanisms of flight. Of these, one group is invertebrate—the insects. Each of three vertebrate classes has developed flight. Among the reptiles there are the pterosaurs, which have been extinct now for millions of years. Among the mammals there are the bats. Finally, and most impressive of all, there is virtually the entire class of birds.

As long as birds easily and, apparently, effortlessly cleave the atmosphere, human beings could not help but dream of doing so themselves.

It is not easy, however, to keep a massive object suspended against gravity through the support of a medium as rarefied as air, and it only works because flying animals are comparatively small in size.

Insects are the smallest of the different kinds of fliers, for they have a less efficient breathing mechanism than vertebrates do and could not produce the energy needed to keep more than a very small mass in rapid motion. The most massive insect in existence is the Goliath beetle, which can weigh as much as 100 grams.*

Vertebrate fliers, with efficient lungs and with wings that have internal body stiffening, light but strong, can be considerably larger than that. Though there are bats that are smaller than the largest insects, the largest bats (such as some of the fruit bats in Indonesia) weigh as much as 900 grams.

Some birds are larger still. The most massive flying birds are Kori bustards, at least one of which was reported to weigh 24 kilograms. Some of the extinct pterosaurs may have approached this mass, and fossil remnants of a very large one that may have exceeded this mass have recently been reported, but there is some question as to whether the large pterosaurs actually flew or merely glided.

It is probably safe to say that in all the history of biological evolution on Earth it is doubtful if flight has ever been developed for any animal weighing more than 30 kilograms at the very most.

For this reason, there doesn't seem any chance of human flight in any fashion resembling that of birds, for even a small human adult is likely to weigh more than 35 kilograms, and an average American male adult will weigh twice that.

Until modern times, however, people did not take this into account. When the ancient Greeks told the myth of the great inventor Daedalus, they had him escape from Crete by constructing a pair of wings for himself (and another pair for his son, Icarus). The wings consisted of a light framework to which feathers were attached by wax. It was the feathers which were thought to confer the gift of flight.

Artists who felt it necessary to indicate the supernatural ability of angels to fly by giving them wings gave them long bird wings without depriving them of arms, apparently not quite grasping the fact

---

* A gram (1/1000 of a kilogram) is 0.035 ounces in common American measure. so that 85 grams equals 3 ounces.

that they were making angels six-limbed. (On the other hand, fairies, particularly if imagined to be small, were often given the gauzy wings of insects, while demons were given the leathery wings of bats.)

Animals far more massive than men were imagined to be capable of flight. There was the flying horse, Pegasus, in the Greek myths; there were flying dragons, huge reptiles oddly reminiscent of the pterosaurs, held aloft by batwings that were often pictured as ridiculously tiny. One famous tale in *The Thousand and One Nights* tells of the "roc," a flying bird so huge that its egg was as large as a small house, and so powerful that it could lift an elephant in one set of talons, a rhinoceros in the other, and fly off with them. The roc is thought to have been inspired by the aepyornis of Madagascar (not too long extinct), which was the largest bird that ever lived. It stood up to 3 meters tall and weighed up to 450 kilograms—but was utterly incapable of flight.

Even inanimate objects were dreamed of as capable of flight; witness the flying carpets of legend and fancy


## HUMAN FLOATING

To be sure, human beings could make objects that would fly, after a fashion. Since it was known that the wind would whirl light, flat objects aloft (light so that not much force was required, flat so that as much wind as possible was caught), it was bound to occur to people to stretch paper, or the equivalent, across a flimsy wooden framework and attach to it a long cord by which it could be held. In that way, one has a "kite."

The use of kites goes back to prehistoric times in eastern Asia. In Western tradition, a kite is supposed to have been invented by a Greek philosopher of southern Italy, Archytas of Tarentum (400–350 B.C.).

Kites were used for thousands of years, chiefly for amusement, though practical uses were possible. A kite can hold a lantern aloft as a signal to troops over a wide area. A kite can also carry a light cord across a river, a cord which can be used to pull heavier cords across until a bridge can be started.

The first attempt to use kites for exploring the properties of atmospheric heights came in 1749. In that year, a Scottish astronomer, Alexander Wilson (1714–86), attached thermometers to kites, hoping to measure temperatures at a height.

Much more significant was the work of the colonial American

scholar Benjamin Franklin (1706-90), who in 1752 flew a kite in a gathering thunderstorm and charged a Leyden jar with electricity, proving that lightning was an electrical discharge. In this way, he developed the lightning rod and eliminated much of the terror of the lightning stroke. It was the first victory of science over a natural calamity.

It was not long before human beings were following, in person, on the track of kites.

Two Frenchmen, Joseph Michel Montgolfier (1740-1810) and his younger brother, Jacques Etienne (1745-99), observed the manner in which the smoke of fire caught up light objects and lifted them into the air. Hot air, it seemed to the brothers, was lighter than cold air, and floated on cold air as wood floated on water. It seemed natural to suppose that if a light bag was held, opening downward, over a fire, it would fill with hot gases and be carried upward until such time as the gases cooled and no longer had levitating power.

On June 5, 1783, in the marketplace of their hometown, the brothers filled a large linen bag with hot air. It lifted 450 meters upward and floated a distance of nearly 2.5 kilometers in ten minutes. This was the first "balloon" (from an Italian word for a large ball, or sphere).

It seemed obvious that if one suspended a light gondola from the bottom of the bag, the gondola and (within reason) any contents it contained would also be lifted. On September 19, 1783, the Montgolfier brothers went to Versailles, the seat of the French government, and there, before a notable audience, they demonstrated a larger balloon with a gondola within which were a sheep, a rooster, and a duck—the first land animals to become aeronauts.

The first human beings were carried along on November 21, 1783. Before 300,000 fascinated spectators, among whom was Benjamin Franklin, a balloon was filled with hot air produced by a large fire, and it then managed a flight of 19 kilometers in twenty-three minutes.

Hot air is an inefficient lifting medium, however. It is not much lighter than cold air, and so it doesn't have much lift, and what little it has it loses as it cools.

The gas hydrogen had, however, been produced and studied by Henry Cavendish in 1766. It was the lightest gas known (either then or now), with a density only 1/14 that of air. Hydrogen has, therefore, at least three times the lift of hot air, and has it even when it is no warmer than the air about it, so that it requires no heating and loses nothing by cooling.

When the French physicist Jacques Alexandre Charles (1746-

1823) heard of the Montgolfier balloons, he suggested the use of hydrogen, and that suggestion was followed before the end of the year. In December 1783, Charles himself, together with a friend, were the first to soar over the surface of Earth in a hydrogen balloon. They drifted for 24 kilometers and landed in a rural village, where the villagers, horrified at the apparition of what must have seemed like a dragon from the sky, bravely attacked the balloon with pitchforks and "killed" it.

Ballooning caught on as a sport that combined the brand-new sensation of drifting through the air and viewing Earth's surface from a height with that spice of danger that adds enormously to the pleasure of some people.

The value of the balloon as an instrument of exploration was also immediately apparent. In 1784, within a year of the innovation of the balloon, an American physician, Joe Jeffries (1745-1819), made a balloon ascent over London, carrying with him a barometer and other instruments, plus an arrangement to collect air at various heights.

That same year, a French aeronaut, Jean Pierre Blanchard (1750-1809), invented the parachute. This is a device in which a man is suspended from a large, curving strip of linen, silk, or other material which catches air within itself greatly increasing air resistance and slowing the rate of fall, in consequence, to a safe level. Blanchard himself made the first parachute jump in 1784 and survived handily.

It was quickly shown that the balloon could rival the climbing of mountains as a way of studying the upper air. In 1804, the French chemist Joseph Louis Gay-Lussac (1778-1850) ascended some 7 kilometers above Earth's surface and brought down samples of air. This is nearly one and a half times the height of Mont Blanc. What's more, the balloon ascent was far quicker, less effortful, and, on the whole, less dangerous than the ascent of Mont Blanc would have been. Again, the balloon could study the air anywhere, whereas at Mont Blanc the air could be studied only there.

The balloon was the first, but not the only, device that carried human beings into the air. As the nineteenth century progressed, a number of people experimented with "gliders," which were essentially kites that were large enough and sturdy enough to support the weight of a human being.

Outstanding in this respect was the German aeronautical engineer Otto Lilienthal (1848-96), who began designing gliders in 1877. In 1891, he launched himself on his first glider flight and was eminently successful.

Gliding is not as passive an experience as ballooning is. The person on the glider could, within limits, manipulate the wings and controls of the device in such a way as to direct its flight. Gliding became a danger-sport craze of the 1890s as ballooning had been a century earlier. Lilienthal launched himself into the air over two thousand times until, in 1896, while testing a new rudder design, he crashed and died of his injuries.

## HUMAN FLYING

How could one truly control an aeronautical device? Suppose, for instance, the gondola of a balloon carried a steam engine that could turn a propeller. Would it not send the balloon through the air as the equivalent device successfully sent a ship through the water? One would then have an "airship."

It was not quite that simple. A steam engine is a heavy device, and it would have to burn large quantities of heavy wood or coal. Lifting all of that would be a problem. In 1876, however, the German inventor Nikolaus August Otto (1832–91) built the first practical internal-combustion engine. This could be made lighter and more powerful than a steam engine. The notion of an airship came further within the realm of possibility.

A balloon itself, inevitably spherical in shape, is not aerodynamically efficient. Pushing it through the air would involve unnecessary air resistance. It occurred to a German inventor, Ferdinand von Zeppelin (1838–1917), to place a balloon, or balloons, within a metal framework that would force the whole into a cigar shape that could cleave the air with minimum resistance.

Obviously, a metal would be needed that was at once light and strong. For that purpose, aluminum was satisfactory. It had been produced in quantity for the first time, in 1886, by a method devised independently by the American chemist Charles Martin Hall (1863–1914) and the French chemist Paul Louis Héroult (1863–1914).

On July 2, 1900, one of Zeppelin's aluminum cigars rose in the air. Under it was a gondola containing an internal-combustion engine that turned propellers. The airship flew successfully, and though it was somewhat damaged on landing, this represents the first successful *powered* flight by human beings. Such an airship was also called "a dirigible balloon" (that is, one that could be directed) or "dirigible" for short; and in honor of its inventor, it was also called a "zeppelin."

For nearly forty years, dirigibles were a beautiful and impressive, though not common, feature of the skies. The most successful of them was the German *Graf Zeppelin,* named for the inventor. However, the huge objects were vulnerable to storms and accidents, and the last straw came in 1937, when the German dirigible *Hindenburg* caught fire and was destroyed. With that, the large dirigible, as passenger liner of the sky, passed from the scene.

Even while the first successful dirigible was being constructed, there were also attempts to construct what we might call a "dirigible glider" by mounting an internal-combustion engine upon one.

The American astronomer Samuel Pierpont Langley (1834–1906) labored for years at this task and, between 1897 and 1903, built three such devices but failed, though each time by a narrower margin.

Two brothers, Wilbur Wright (1867–1912) and Orville Wright (1871–1948), were bicycle repairers by trade and gliding enthusiasts by hobby, and they too labored to build a powered glider. They followed Langley's work, made shrewd corrections in design, and invented a device that was a predecessor to the ailerons, or movable wing edges that enable a pilot to control the device more easily. In addition, they built a wind tunnel in which to test their models, designed new engines of unprecedentedly low ratios of weight to power, and, over a period of eight years, spent no more than $1,000.

On December 17, 1903, at Kitty Hawk, North Carolina, Orville Wright piloted the first flight of a powered device that was heavier than air. Such a device is popularly called an "airplane," and what holds it up is the rapid movement of air beneath its wings.

The first flight endured for almost a minute and covered only 260 meters, but it led to all the rest. A dozen years later, airplanes were fighting duels in the air above Europe and lending the one touch of glamour to the stupidity and misery of World War I. In 1927, the American aeronaut Charles Augustus Lindbergh (1902–74) flew nonstop across the Atlantic Ocean from New York to Paris, and with that the airplane truly came of age.


## THE UPPER ATMOSPHERE

Until early modern times, it was taken for granted that the air above the surface of Earth stretched upward indefinitely and filled the universe, reaching to the moon, the planets, and the starry limits.

To be sure, if the heavenly bodies passed through air as they revolved about Earth, air resistance would cause them to lose energy, spiral inward rapidly, and collide with Earth. The fact that this did not happen was proof enough that the heavenly bodies did *not* pass through air. This, however, was not plain until the end of the 1600s, when the laws of motion first came to be understood.

When early writers imagined trips to the moon and beyond, therefore, they required only some device for traveling through the air.

The earliest fiction by known authors concerning trips to the moon were written by the Greek satirist Lucian of Samosata (120–180) about A.D. 165. In his tale *Icaromenippus,* a philosopher is described as flying to the moon by using the wing of a vulture on one side and the wing of an eagle on the other. In a later book, *A True History,* Lucian describes a ship as being caught up in a whirlwind and, after a week-long journey, reaching the moon.

In 1532, the Italian poet Ludovico Ariosto (1474–1533) wrote *Orlando Furioso,* in which his hero, Orlando, reaches the moon by means of the same chariot that had carried the prophet Elijah to heaven.

The first tale involving a trip to the moon that was written in English was published posthumously in 1638 by Francis Godwin (1562–1633). It was called *The Man in the Moone,* and in it the hero is pulled to the moon in a vehicle hitched to large swanlike birds which were said to migrate regularly to the moon.

Five years later, however, came the turning point. In 1643, the Italian physicist Evangelista Torricelli (1608–47), showed that the weight of air was sufficient to support a column of mercury 76 centimeters * high. The weight of a column of air, then, had to be equal to that of such a column of mercury. The density of mercury is 11,450 times that of air at sea level, and so if air were at a uniform density all the way up, the column of air required to balance 76 centimeters of mercury would be 8.7 kilometers high.

This was the first indication that the atmosphere was a local phenomenon only—that it represented a relatively thin layer about Earth and that beyond it was vacuum.

To be sure, the atmosphere is not of uniform density throughout. The atmosphere at sea level is under the weight of all the layers of air above, and that compresses it to its particular density. As one moves higher in the air, more and more of the atmosphere lies below, and there is less and less above to apply pressure. Air pressure falls with height, and so does density.

This was first demonstrated by actual observation in 1646, when the French mathematician Blaise Pascal (1623–62) sent his brother-

---

* A centimeter is 1/100 of a meter and is equal to about 2/5 of an inch in common American measure, so that 5 centimeters equals 2 inches.

in-law up a local mountainside with two barometers. The steady drop of air pressure with height was recorded.

As a result of this drop in density, the atmosphere is spread out over a greater volume and its height is considerably more than the 8.7 kilometers it would be under conditions of constant density. The drop in density is such, however, that in any practical sense, the atmosphere remains an extremely thin layer over Earth's surface. By the time it is 8 kilometers high or so, it is so thin that it can barely support life. It is for this reason that in climbing high Himalayan peaks, oxygen cylinders are a necessity.

Even this knowledge of the earthbound nature of the atmosphere, and the realization that between the astronomical bodies lie vast stretches of vacuum, did not stop science fiction writers altogether.

Once the balloon was invented, that seemed the proper method for reaching great heights. Whirlwinds, birds, and wings vanished. As late as 1835, Edgar Allan Poe (1809–49) published *Hans Pfaall,* a story in which the hero gets to the moon by balloon. Poe appreciates the decreasing density of the air but gets around that by imagining a device called a "condenser" which compresses the air about the balloon. In real life, of course, there are no such "condensers" even today, and not much chance of one in the future, so that the height limit to which balloons can climb lies far below the moon.

For one thing, there is the matter of breathing. In 1875, three French aeronauts rose to a height of some 9.6 kilometers in a balloon, about 800 meters higher than the peak of Mt. Everest. Unfortunately, two of the aeronauts did not survive the grueling experience. The one survivor, Gaston Tissandier (1835–99), was able to describe the symptoms of air deficiency, and he is therefore considered the father of "aviation medicine."

Airplanes are less well equipped for high flights than balloons are. As they rise and the air grows thinner, planes must go faster to maintain the necessary lift, and that limits the height to which they can go. On the whole, if great heights are to be reached, it is the balloon that must be depended upon.

But if balloons were to rise higher than Tissandier's mark of 9.6 kilometers, it was quite clear that open gondolas had to go. Sealed cabins were required that would keep the air pressure within at normal level. This meant that carbon dioxide would have to be absorbed as produced, while oxygen was renewed.

The problem was not as difficult as that of sealing normal air within a submarine. In a submarine, the external pressure can rise to many atmospheres. In a balloon gondola, the external pressure cannot fall to less than zero atmospheres. The differential in a submarine can therefore be many atmospheres, while in a balloon gondola it is always less than one.

Then, too, better material was found for the manufacture of balloons. Silk is fairly heavy in the thickness and layers required to build an efficient balloon large enough to rise to great heights. In addition, silk, however tightly woven, is rather porous to hydrogen. The balloon therefore slowly loses hydrogen and absorbs air instead, so that the lift gradually decreases. As the twentieth century wore on, however, plastics could be used for the manufacture of balloons. This could be made thinner and lighter than silk, and yet be strong enough for the purpose, and be less porous than silk, too.

One danger in balloons (and in airships, too) rested in the inflammability of hydrogen. Fires can begin as the result of accidental sparks, or electrical discharges, and destroy the vehicle and its passengers. A hydrogen fire destroyed the *Hindenburg,* for instance.

In 1895, however, the Scottish chemist William Ramsey (1852–1916) discovered the rare gas helium, second only to hydrogen in low density. The density of helium is twice that of hydrogen, but even so, helium has more than nine-tenths the lifting power of hydrogen. Its slight inferiority to hydrogen in lifting power is much more than made up for by the fact that helium is absolutely inert and will not burn. Nor does it leak through the substance of the balloon as readily as hydrogen does.

In 1931, Auguste Piccard, who was later to design the bathyscaphe, used a plastic balloon with a sealed gondola and rose 15.8 kilometers into the air—nearly twice as high as Mt. Everest—in an eighteen-hour flight.

This was the first time that human beings, or, indeed, any living things (except, perhaps, wind-blown microscopic spores), had risen beyond the highest clouds.

Those changes that characterize weather—cloud formation, precipitation, turbulence—take place in the lowest level of the atmosphere. This lowest region was therefore named the "troposphere" ("sphere of change") in 1908 by the French meteorologist Léon Teisserenc de Bort (1855–1913).

The upper boundary of the troposphere Teisserenc de Bort called the "tropopause." The altitude of the tropopause varies with latitude, from a height of about 16 kilometers at the equator to only 8 kilometers at the poles.

Above the tropopause, Teisserenc de Bort believed, the thin atmosphere lacked turbulence and change, and existed in quiet layers. He called it the "stratosphere" ("sphere of layers"). It was into the strat-osphere that Piccard successfully carried through the first human penetration.

Balloons have gone higher still. In 1961, a balloon with two American naval officers aboard reached a height of 34.668 kilometers, and in 1966, an unofficial record of 37.7 kilometers was set. Unmanned balloons have reached heights of nearly 47 kilometers, over five times the height of Mt. Everest.

## RADIO WAVES AND ELECTRONS

It doesn't seem likely that balloons, or any devices that require air to maintain a height above Earth's surface, are going to go substantially higher than the records achieved in the last two decades. There are, however, indirect ways of studying the atmosphere at even greater heights.

On December 12, 1901, the Italian electrical engineer Guglielmo Marconi (1874–1937) succeeded in sending radio signals from England to Newfoundland across the Atlantic Ocean. This is usually taken as representing the invention of radio. Such radio signals are transmitted by the use of radio waves, similar in nature to light waves, but a million or so times longer.

Like light waves, radio waves travel in straight lines, and so a radio-wave transmission should not be detectable beyond the horizon. Nevertheless, Marconi's signals traveled from England to Newfoundland around the curve of Earth.

A British-American electrical engineer, Arthur Edwin Kennelly (1861–1939), and an English electrical engineer, Oliver Heaviside (1850–1925), independently suggested in 1902 that radio waves would be reflected by ions (electrically charged atom fragments) and that there must be a layer of ions high in the atmosphere that reflected radio waves. (This came to be called the "Kennelly-Heaviside layer.") Bouncing between the Kennelly-Heaviside layer and the ground, radio signals could travel around the curve of Earth's globe.

The English physicist Edward Victor Appleton (1902–65), studying the manner in which beams of radio waves interfered with each other, produced convincing evidence, in 1922, that the theoretical suggestions of Kennelly and Heaviside were accurate and that there was indeed an ion-rich layer in the upper atmosphere. By 1924, he was able to show that the Kennelly-Heaviside layer was some 95 kilometers above the surface of Earth. He also advanced evidence for the existence of still higher ion-rich regions ("Appleton layers") and, in 1926, showed that some of these were as much as 240 kilometers high.

The portion of the atmosphere lying between heights of 50 and 300 kilometers above the surface of Earth is therefore called the "ionosphere."

The density of the gases in the ionosphere is only in the range of a billionth that of the atmosphere at sea level, but this is still much greater than the ultra-tenuous vapors present in deep space, and is dense enough to produce easily perceptible phenomena.

Earth, for instance, is constantly being bombarded with vast numbers of tiny bits of matter (probably the residue of long-disintegrated comets) entering the atmosphere as "meteoroids," which are the size of pinheads or less. As they travel rapidly through the atmosphere at speeds of 20 kilometers per second and more, they compress and heat the gases ahead of them, reaching temperatures at which they glow visibly and can be detected from Earth by the unaided eye as "meteors" or "shooting stars." These meteors glow and burn while they are still in the ionosphere. They not only demonstrate the existence of the thin gases of that region that serve as protection against this multitudinous pinhead bombardment, but also offer us data that serve to feed us information about those gases.

There are clearly visible phenomena that exist at heights greater still.

These depend on the fact that Earth is itself a magnet, a fact that was first clearly demonstrated in 1600 by the English physicist William Gilbert (1544–1603). This means that Earth is surrounded by magnetic lines of force, just as any magnet would be.

A moving object which is electrically charged interacts with magnetic lines of force, and must expend energy in order to cross them. It requires less energy for such charged objects to move along lines of force than to cross them (just as it takes us less energy to walk along a level floor than to move up or down flights of stairs).

All matter is composed of atoms, which are, in turn, composed, in the main, of three varieties of "subatomic particles." Of these subatomic particles, two carry electric charges. The "proton" carries a positive electric charge and the "electron" a negative one.

The electron, first identified and studied in 1897 by the English physicist Joseph John Thomson (1856–1940), is much the lighter of the two, possessing only $1/1,837$ the mass of the hydrogen atom (the smallest of all atoms). The electron, because of its lightness, is particularly easily deflected by the magnetic lines of force and tends to run along them in tight spirals.

Earth's magnetic lines of force run from the north magnetic pole to the south magnetic pole, bellying far out into space in between. The north magnetic pole is located in northernmost Canada and the south magnetic pole at the edge of Antarctica. The magnetic lines of force curve downward toward Earth in the two polar regions, therefore, and the electrons trapped in those lines of force move downward as well.

As the electrons move downward, they reach the upper reaches of the atmosphere, where they collide with atoms. Some of the energy of collision is converted into flashes of light. The polar regions are therefore the scenes of luminous displays visible during the long nights.

These are the "aurorae," from the Latin word for "dawn," since travelers moving northward saw the dim light on the northern horizon and thought it looked like the dawn breaking in the wrong direction. The "Northern Lights" of the Arctic are the "Aurora Borealis" ("northern dawn") and the "Southern Lights" of the Antarctic are the "Aurora Australis" ("southern dawn").

The shifting colored streamers of light make the aurorae a beautiful sight, and the height at which they are visible stretches from about 100 kilometers upward to extremes of 1,000 or even 2,000 kilometers. At those extreme heights, there is still a sufficient scattering of atmospheric atoms for the speeding electrons to collide with and produce a visible effect.

The region beyond the ionosphere that is capable of producing auroral effects is called the "exosphere." The exosphere fades imperceptibly into the vacuum of interplanetary space. (This vacuum is not *true* emptiness. Even very far from any planet there is an exceedingly sparse scattering of various kinds of atoms—but not enough of them to produce a visible aurora.)

We can say, then, that Earth's atmosphere, judged by its perceptible effects, extends upward from Earth's surface for some 2,000 kilometers. By the most extreme accounting, then, Earth's atmosphere is a local phenomenon that extends only about 1/200 of the distance to the moon. What's more, unmanned balloons have risen upward less than 1/8,000 of the distance to the moon, and manned balloons only 1/10,000 of that distance.

# 8
# Beyond the Balloon

## ACTION AND REACTION

It might have seemed, to those who saw nothing beyond the balloon, that humanity's horizon upward was sharply limited in the upward direction, and perhaps forever so.

Yet suppose we consider locomotion.

When we walk, we progress forward by pushing backward against the ground with our feet. If you don't see the importance of this push, try to walk on very smooth ice, where, because of lack of friction, your feet cannot get purchase on the surface you are standing on. You will slide this way and that, and will probably fall, but you will make no progress. This is true of all land animals whether they walk as we do, run as horses do, hop as kangaroos do, or slither as snakes do.

Similarly, we progress in the water by pushing backward against the water with our hands and feet, just as fish do with their tails, seals and penguins with their flippers, whales with their flukes, and so on.

The propellers of human devices are mechanical analogues of the flippers of sea creatures and the wings of birds.

As long as we can progress forward *only* by a backward push on the medium we travel on or through, then space presents us with a serious problem. Space is essentially a vacuum and contains nothing to push against.

Yet motion through space is possible, since both the moon and Earth move through space, have been doing so for billions of years, and will, in all likelihood, continue to do so for billions of years more.

In the case of the heavenly bodies, speed is crucial. If the moon were to slow continually in its motion around Earth (as it would if it were traveling through air), it would spiral inward toward Earth and would eventually collide with it. Similarly, a slowing Earth would spiral in toward the sun and would collide with it.

If, somehow, an earthly vehicle were to gain enough speed, it could go into orbit about Earth and remain there indefinitely (if it were high enough to avoid the loss of energy through friction with traces of air in the upper atmosphere). The minimum speed required for moving into orbit is about 8 kilometers per second. A speed of 11.2 kilometers per second ("escape velocity") will allow a vehicle to move away from Earth indefinitely.

If a vehicle merely gained sufficient speed to go into orbit and could do nothing more for lack of a medium to push against, it could do nothing but respond to gravitational fields. It would be in "free fall," moving automatically in some orbit, as the moon does about Earth and Earth does about the sun.

An orbiting vehicle, in free fall about Earth, could still be enormously useful despite its inability to maneuver. It could carry instruments that might tell us much about conditions in space that we could not learn from our surface-bound devices.

The question, then, is whether a vehicle can gain enough speed to go into orbit at a great enough height to remain in orbit for a considerable period of time.

As long as we must use devices that progress by pushing against a medium, the chances would seem to be nil. The fastest device, natural or human-made, that makes such progress is the propeller-driven airplane, and the greatest speed such planes have obtained is about 0.25 kilometers per second. This is only 1/32 of the speed required to put an object into orbit about Earth. Nor is it likely that any further improvement of the propeller-driven airplane would increase that speed significantly.

Is there any way of moving forward without pushing against a medium?

The answer is—yes! There is actually a living organism that uses a means of propulsion fundamentally different from all those I have described so far.

The squid can propel itself forward by expelling a jet of water backward. The backward motion of the jet is balanced by the motion of the squid forward. It might seem as though the jet of water pushes against the rest of the water in order to drive the squid forward, but this is not so. If the squid were in a vacuum (and could survive under such conditions) and if it had water to squirt backward, it would still move forward, even though there was nothing for the squirted water to push against.

Thus, imagine yourself sitting on a frictionless surface—a highly waxed floor, a smoothly iced pond, or something of that sort, something so frictionless that you could not propel yourself by pushing your hands against the material on which you were sitting simply because your hands would slide along uselessly.

Imagine, further, that you have a pile of fairly heavy rocks in your lap. If you throw a rock in a particular direction, you will find yourself sliding slowly in the opposite direction. If you throw a second rock, then a third, then a fourth, all in the same particular direction, your speed will grow greater with each thrown rock. Again, this does not happen because the thrown rock pushes against the air. It would happen even if you were in a protective spacesuit and were sitting surrounded by vacuum. In fact, it would happen more efficiently in a vacuum because there would be no air resistance to get in the way of the motion of either the rock or yourself.

This method of starting from rest and producing motion in two opposite directions is an example of something called "the law of conservation of momentum." This particular example of that law is also called "the law of action and reaction" and was first propounded in a systematic way by Newton in 1687. Still another name for it is "Newton's Third Law of Motion."

(To begin with, the law of action and reaction was purely theoretical as far as space was concerned, since Newton was never in space to test his notion, but it has since been amply tested and you can be sure it holds.)

The principle could be used in airplanes. Suppose the fuel is burned within the engine and the hot exhaust gases are forced rapidly back through a nozzle to form a high-speed "jet" of gas. The backward thrust of that jet pushes the plane forward.

The first to invent a practical jet engine for aircraft was Frank Whittle (1907–    ), a British engineer. He obtained a patent on the essential device in 1930. The Germans first flew a jet plane in August 1939, and the British in May 1941. Soon the Americans had them also, and by the end of World War II, jet planes were being used in combat.

Jet propulsion developed rapidly. Within a few years, jet planes were flying faster than propeller planes and commercial air travel drowned out all other kinds for long-distance travel as very large, very fast jet planes came into use.

On October 14, 1947, the American pilot Charles E. Yeager (1923–      ) flew faster than the speed of sound, something that was publicized as "breaking the sound barrier." He attained a speed of 1,080 kilometers per hour under conditions of air temperature and pressure in which the speed of sound was 1,060 kilometers per hour.

Jet planes have since flown at speeds of up to about 7,250 kilometers per hour, or just about 2 kilometers per second, but even this is only a quarter of the speed necessary to push an object into orbit around Earth.

To gain the necessary speed, it would help if a device could accelerate at great heights where air resistance is low enough not to interfere too much (or to heat the device too dangerously, for that matter). However, a jet plane needs oxygen if its fuel is to burn, and that oxygen must be gathered from the atmosphere. If the plane were high enough to accelerate easily without undue fear of resistance and heat, there would not be enough oxygen pressure in the atmosphere outside to support the burning of the fuel.

Even jet planes don't seem to be the answer to the problem of exploring space.

## ROCKETS IN EMBRYO

What is needed, then, is a device that uses the jet principle and burns something that doesn't need atmospheric oxygen to burn.

The first substance found to burn without atmospheric oxygen was a mixture of potassium nitrate (saltpeter), sulfur, and charcoal. The sulfur and the charcoal are inflammable; that is, they will combine rapidly with oxygen. The oxygen can be obtained from the potassium nitrate, where it is in chemical combination with potassium and nitrogen. When the three substances are mixed in the proper proportions and heated, the chemical combination is very rapid and leads to the production of a large volume of vapor.

The mixture is "gunpowder." If the mixture is penned up in an enclosed place, the rapid evolution of gas takes place anyway, on heating, since no atmospheric oxygen is needed. The pressure of the vapor that is formed is likely to burst the container with a loud report of explosion. Gunpowder was the first "explosive" discovered.

Gunpowder was first discovered in China during the Middle Ages, and knowledge of it reached Europe in the 1200s. For some six centuries thereafter it remained the only explosive known.

If gunpowder is placed in a cylinder with a small opening at one of the ends, then when the gunpowder is ignited, the vapors push forcibly out of that small opening and the cylinder itself moves rapidly in the other direction. The result is a "rocket."

Rockets first appeared in warfare in 1232, when the Chinese used them against attacking Mongols. For centuries thereafter they were used sporadically. Gradually, they were made more massive and powerful and the rocket principle sent masses of speeding projectiles at an enemy force. In the 1790s, the Indian armies of Tipu Sahib (1751-99) used them against the British.

A British artillery officer, William Congreve (1772-1828), having experienced these rockets, set about improving them further. In 1805, he built a rocket a meter long, with a stabilizing stick to keep it moving in the aimed direction without tumbling or veering. The stick itself was 5 meters long and the rocket had a range of 1.8 kilometers.

What were called "Congreve rockets" were used, by the British, in bombardments during the Napoleonic wars. They were also used by them in the bombardment of Fort McHenry in Baltimore Harbor during the War of 1812. Francis Scott Key (1779-1843), who observed the bombardment and who wrote the verses now known as "The Star-Spangled Banner" on that occasion, mentioned "the rockets' red glare" with reference to the Congreve rockets.

Another way of delivering masses at high speed on enemy forces was conventional artillery. Quantities of gunpowder were placed in a cylinder closed at one end (cannon) and a tight-fitting ball of stone or metal was placed before it. The explosion of the gunpowder drove the "cannonball" out the open end of the cannon and at the enemy.

Cannons were heavy, clumsy, hard to manufacture, and the balls they hurled moved at their fastest when they left the mouth of the cylinder. The firing also involved a powerful recoil. In comparison, rockets were lighter, easier to handle and manufacture, and moved faster and faster as long as the gunpowder lasted. What's more, there was no recoil.

Nevertheless, advances in artillery during the early nineteenth century made cannon fire much more accurate than rocket fire. What's more, artillery delivered heavier masses. Congreve rockets therefore grew obsolete.

Rocketry did not die out altogether, however, and rockets continued to be used for minor functions in warfare. During World War II, rockets began to be important again. The American army, for instance, developed the "bazooka." It worked on the principle of the blowpipe, along which a poisoned dart can be forced by a puff of breath. The bazooka was a 1.5-meter-long cylinder along which a lighted rocket would make its way and be aimed, effectively, against oncoming tanks. The bazooka could be aimed by a single soldier. No piece of artillery of equivalent power could be carried by one soldier; nor could an artillery piece be counted on to be recoil-free.

The Soviet army made effective use of serried ranks of rocket launchers, a weapon they called "Katyusha."

Rockets were not conceived as weapons of war only. Rockets, using gunpowder, were, after all, an example of a moving object that could perform in a vacuum, maneuvering there, and they could therefore be visualized as carrying instruments, and even human beings themselves, beyond the atmosphere.

The first occasion on which this was described was in a science fiction novel written by none other than the French soldier, duelist, and poet the long-nosed Cyrano de Bergerac (1619–55). He wrote a science fiction romance, *Voyages to the Moon and the Sun,* which was published posthumously in 1657. In it he has his hero plan a voyage to the moon by various artifices, most of them fanciful and useless. One of them, however, involved the use of rockets. The hero of the tale even used the rockets, which, however, were not sufficient to carry him to the moon. The feat was carried through to a successful conclusion by one of the other methods.

Cyrano's leap of imagination occurred thirty years before Newton had placed rocketry on a firm theoretical foundation with his Third Law of Motion. Cyrano's rocketry was not, however, picked up by other science fiction writers.

In 1827, the American educator George Tucker (1775–1861) published *A Voyage to the Moon,* in which he made use of a gravity-neutralizing substance for the purpose. This same device was used in 1901 by the British writer H G. Wells (1866–1946). Unfortunately, antigravity is a fantasy even today. It is not only beyond the bounds of present technology, but it may even be theoretically impossible (although it is always dangerous to make such a statement).

In 1835, as I stated earlier, Edgar Allan Poe used a balloon to get his hero to the moon. This is, of course, utterly impossible.

In 1865, the French writer Jules Verne (1828–1905) wrote *From the Earth to the Moon,* in which his heroes reached the moon after having been shot out of a giant cannon.

This could indeed work, in the sense that an object could conceivably be fired by so huge a charge of explosive as to emerge from the mouth of the cannon at a speed greater than escape velocity. The object would then be hurled through the atmosphere (provided its speed was sufficiently above escape velocity so that the slowing effect of air resistance would not bring that speed to a figure below the crucial amount) and into space, through which it would travel indefinitely. Its path would be curved, under the influence of Earth's gravity, but if that is allowed for, and if it is properly aimed, its motion could carry it to the moon.

The trouble is, however, that the full speed of the object is gained over the length of the cannon, and that rapid acceleration would smash and kill any human occupants of the vehicle. A rocket, on the other hand, gains its speeds over many kilometers, doing so for as long as its fuel keeps burning and the developing gas jets backward. The acceleration is therefore lower, and while it takes longer for a rocket to gain escape velocity than for a cannon-fired object to do so, the former has the advantage, at least, of keeping its crew alive.

In 1869, the American writer Edward Everett Hale (1822–1909) wrote a tale entitled *The Brick Moon* in which an unfinished space vehicle is accidentally placed in orbit about Earth when it is hurled upward by the force of two giant flywheels. This is essentially equivalent to being fired out of a cannon, and the same objection can be raised.

The only rocket to appear in nineteenth-century science fiction was in an obscure book, *Voyage to Venus,* published in 1865 by a French writer, Achille Eyraud, who, however, got the working principle of the rocket wrong. He thought the rocket exhaust could be caught, brought back, and used over and over again.

Inventors did rather better with the rocket principle than science fiction writers did, on the whole.

About the time Congreve was producing his rockets, an Italian fireworks manufacturer, Claude Ruggieri, was sending up mice and rats in containers attached to rockets and bringing them safely back to Earth by parachute. The story is that he planned to use a cluster of large rockets that would have enough power to send a sheep aloft, or possibly a boy, but that the police stopped that.

Then there was the case of a Russian explosives maker, Nikolai Ivanovich Kibalchich, who manufactured the bombs used to assassinate Czar Alexander II of Russia in 1881.

Kibalchich was tried (along with others), convicted, and executed. While he was awaiting execution, however, he worked out a plan for a rocket-powered aircraft, wrote it out in detail, and presented it to the prison authorities. The authorities, unwilling to deal with the work of a revolutionary, placed it under lock and key, and there it remained until the Russian Revolution.

Even bolder was a German law student, Hermann Ganswindt (1856–1934), who dealt with all sorts of fanciful vehicles such as airships and helicopters. As early as 1891, he was thinking of rocket-powered vehicles and, for the first time in history, specifically tried to work out the design of a rocket-powered device that would attain escape velocity. He was not a scientist, however, and his design was entirely impractical.

## ROCKETS IN EXPERIMENT

The first person to interest himself in rocket flight into space in a truly scientific manner was a Russian physicist, Konstantin Tsiolkovsky (1857–1935).

Handicapped by almost total deafness from a streptococcus infection at the age of nine, and by the scientific backwardness of czarist Russia, Tsiolkovsky nevertheless educated himself to the point of being able to write scholarly papers on chemistry and physics.

In 1895, Tsiolkovsky mentioned rocket-powered spaceflight for the first time in his papers.

By that time, gunpowder was no longer the only known explosive. New ones, such as nitrocellulose and nitroglycerine, were far more powerful. It was clear that gunpowder was not powerful enough to force a rocket to escape velocity, and Ganswindt, for instance, imagined the use of dynamite for the purpose. (Dynamite is nitroglycerine which has been mixed with inert material to make it safe to handle.)

Tsiolkovsky, however, did not wish to use ordinary explosives, considering them too unpredictable in behavior. He wanted an easily controlled development of energy. In 1898, he worked out, mathematically, the effect of the speed of rocket exhaust on its performance and decided that what was needed was a liquid fuel such as kerosene. If this was mixed, little by little, with liquid oxygen, and the mixture burned, then escape velocity might be attained. Naturally, the spaceship would have to carry both the kerosene and the liquid oxygen so that it would be independent of the atmosphere.

In 1903, he began a series of articles for an aviation magazine in which he went into the theory of rocketry thoroughly. Then and later, Tsiolkovsky wrote of spacesuits, artificial satellites, space stations, and the colonization of the solar system. In later life, he wrote a science fiction novel, *Outside the Earth*, in which he presented his theories for those who would rather read adventures than equations.

In the United States, an American physicist, Robert Hutchings Goddard (1882–1945), had grown interested in rocketry while still a teenager. By 1914 he had obtained two patents involving rocket apparatus. and in 1919 he published a small book. entitled *A Method of Reaching Extreme Altitudes,* in which he discussed the possibility of rocket-powered spaceflight.

In all this he had been anticipated by Tsiolkovsky (whose work in Russia was not really known to the rest of the world). but Goddard then took the crucial step of actually building rockets that would put theory into practice. He experimented first with gunpowder rockets. but in 1923 he began to work with gasoline and liquid oxygen and became the first person in the world actually to construct liquid-fuel rockets.

On March 16. 1926. Goddard fired his first liquid-fuel rocket in Auburn. Massachusetts. His wife took a picture of him standing next to it before it was launched. It was about 1 1/4 meters high and 15 centimeters in diameter. and was held in a frame like a child's jungle gym. After it was fired. it traveled a horizontal distance of 56 meters in 2.5 seconds. reaching a height of 12 meters and attaining a top speed of about 95 kilometers per hour.

This was no more impressive than the Wright brothers' first plane flight at Kitty Hawk a quarter-century earlier. but its significance was equally great.

Goddard managed to get a few thousand dollars from the Smithsonian Institution. and in July 1929 he sent up a larger rocket near Worcester. Massachusetts. It went faster and higher than the first. More important. it carried a barometer. a thermometer. and a small camera to photograph their readings. It was the first instrument-carrying rocket.

This rocket stirred Goddard's neighbors to alarm, and the police ordered him to cease his experiments, since they disturbed the peace. What's more. the *New York Times* berated him at this point. as it had berated Langley earlier. The *Times* editorial writer was convinced that a rocket exhaust had to push against a surrounding medium and therefore could not work in the vacuum of space. From this position of utter ignorance. the editorialist dared to lecture Goddard. a thoroughly skilled physicist. on the basic elements of physics.

Fortunately. Lindbergh. who had just made his epochal flight across the Atlantic. interested himself in Goddard's work. He persuaded a philanthropist. Daniel Guggenheim (1856–1930). to award Goddard a grant of $50,000. With this, Goddard set up an experimental station in a lonely spot in New Mexico where he could work in peace.

In New Mexico. he built larger rockets, and from 1930 to 1935 he launched some that attained speeds of up to 885 kilometers per hour and heights of 2.5 kilometers.

He developed many of the ideas that later became standard in rocketry. He designed combustion chambers of an appropriate shape and burned gasoline with oxygen in such a way that the cold liquids cooled the chamber walls before burning. He also worked out systems for steering a rocket in flight by using a rudderlike device to deflect the gaseous exhaust, with gyroscopes to keep the rocket headed in the proper direction once its course was deflected.

Then, too, he had the notion of using a rocket to carry a smaller rocket aloft. When the first rocket had reached its peak height, it would be loosened and the smaller rocket would begin to burn its fuel and to climb from the height and speed it had already gained. It might be carrying a third still-smaller rocket. Such a "multistage rocket" could climb far higher than a single-stage rocket with the same total fuel, since in the multistage it is only the smaller objects that are sent upward and the mass of the initial rockets is discarded as unnecessary dead weight.

All in all, Goddard accumulated 214 patents on rocketry.

Goddard, however, worked in isolation and was unsupported by the government. During World War II, he worked on several minor projects, and the bazooka was derived from one of his inventions. On the whole, his work did not cause the world to catch fire over the idea of rocketry, any more than Tsiolkovsky's work did.

More important in this way was Hermann Julius Oberth (1894–      ), an Austro-Hungarian physicist born in what is now Rumania. Without knowing of Tsiolkovsky and Goddard, he repeated much of their theoretical work. He attempted to get a Ph.D. with a dissertation based on rocket design, but it was rejected. He published it at his own expense in 1934 as *The Rocket into Interplanetary Space*, and it was this that finally caught the attention of Europeans generally, so that rocket societies were formed and experimentation went into high gear.

## ROCKETS IN ACTION

Thereafter, it was in Germany in particular that the art of rocketry moved forward. The German Rocket Society was founded in 1927, and one of its founding members was the German engineer Willy Ley (1906–69), who wrote an acclaimed book on rocketry while he was still a teenager. Soon following were Walter Robert Dornberger (1895–1980) and Wernher von Braun (1912–77). They began to fire rockets, the first to do so after Goddard.

When Adolf Hitler (1889–1945) became dictator of Germany in

1933, he saw in rockets a potential war weapon of great importance. This meant that government money poured into German rocket research. In 1937, rocket research was shifted to Peenemünde, an isolated area on the Baltic coast. Willy Ley had left Germany soon after Hitler's accession, but von Braun and Dornberger remained and continued to conduct rocket experiments which concentrated more and more on war-related advances.

A pilotless jet plane was developed, for instance, which came into use in 1944. It was a guided missile, a rocket that carried an explosive warhead. Once launched, its course could not be corrected any more than that of a cannonball could. The Germans called it V-1, where V stands for *Vergeltundswaffen,* meaning "vengeance weapon," because by the time it was put into use, Germany was being thoroughly battered by British and American bombers.

The V-1 was about 7 3/4 meters long and had a wingspan of about 5 1/2 meters. It carried a tonne * of high explosive, had a range of about 240 kilometers, and sped along at 580 kilometers per hour, just under the speed of sound.

The V-1 was not really a spaceship, because though it carried fuel it did not carry oxygen. It had to make use of the oxygen of the air; its engines would not work in a vacuum. Between June 13, 1944, and the end of the war, over 8,000 V-1s were launched against Great Britain, mostly against London. An equivalent number were fired against Allied troops in Belgium.

A much more important rocket developed at Peenemünde was the V-2, which was a true spaceship because it carried both fuel (alcohol) and liquid oxygen and if necessary its engines could work in a vacuum.

It was larger than the V-1, being 14 meters long and weighing about 13 tonnes. It carried no more explosive than the V-1 and had very little more range, but it could move at more than the speed of sound, so it was the first supersonic rocket. This meant it could not be shot down by planes or by antiaircraft artillery, since these did not react quickly enough. Nor was its coming heralded by sound.

Beginning on September 6, 1944, 4,300 V-2s were fired against Great Britain and Belgium. Of these, 1,230 hit London, killing 2,511 and wounding 5,869.

Fortunately for the civilized world, these weapons were not developed soon enough to prevent Germany from losing the war.

The British, Americans, and Soviets grasped too late the significance of rockets as war weapons and could not overcome the Ger-

---

* A tonne is equal to 1,000 kilograms, or about 1.1 short tons in common American measure.

man lead by the end of the war. (One of the reasons for this, at least as far as the British and Americans were concerned, was their utter concentration on the development of the nuclear bomb, a much more terrifying weapon that the Germans, in their turn, neglected in favor of rocketry. Of course, the two are now combined, since rocket weapons that have advanced far beyond the V-2 now carry nuclear warheads.)

When the Germans surrendered, the victors each collected as many German rocket experts as possible. Dornberger and von Braun hastened westward to surrender to the Americans, with whom they considered they would be better off. In addition, both the Americans and the Soviets seized such V-2 components as could be found in the rocket factories. The Americans seized enough to put together some hundred V-2 rockets.

In the five years immediately following the German surrender, the Americans fired about seventy V-2 rockets from the White Sands Proving Ground in New Mexico. This taught them a great deal about the handling of large rockets. (Goddard himself died about three months after the German surrender and did not live to see the full vindication of his work.)

With the war over, the V-2 rockets found peacetime uses and were employed in the exploration of the upper atmosphere. In 1949, the United States fired a V-2 rocket to a height of 206 kilometers. In that same year, using a two-stage rocket, a small one on top of a V-2, a height of 400 kilometers was achieved.

This meant that human beings were reaching into the exosphere, where the atmosphere was so rarefied that objects could be put into orbit for extended periods of time.

The V-2 rockets were used up by 1952, but both the United States and the Soviet Union were developing larger and better rockets, and objects in orbit were coming to be well within the limits of the possible.

It was not even something really new in conception. Newton himself had talked about an artificial satellite as a thought experiment nearly three centuries before, and Hale, in his *The Brick Moon,* was the first to write about one in fiction. Tsiolkovsky wrote about an artificial satellite in scientific detail in 1910, and Oberth did the same, independently, in 1923.

To be sure, if an orbiting object was to be useful (aside from merely demonstrating that an orbit could be achieved), it would have to carry instruments and the observations made by those instruments would have to be made available to scientists.

In the case of balloons, there was a "soft landing" and the instruments could be picked up and studied. In the case of rockets, the return was more catastrophic, and if escape velocity was eventually achieved the rockets would not return at all.

It followed that for exploration to be successful, it would be better if the instruments on board the vehicle recorded their observations in such a way as to modify a radio beam from the vehicle. The modified radio beam would thus send the observations to a receiving ground station even as they were made. This is called "telemetering."

Attempts at telemetering were made even before radio came into use. In 1877, telemetering was attempted through electric currents carried by a wire stretching from a balloon to a receiving station. This is not very practical, of course, but it established the principle.

In 1925, the first radio-wave telemetering was attempted for balloon-borne instruments by a Russian scientist named Pyotr A. Molchanov. By the time of the V-2 experiments at White Sands, telemetering was quite advanced and scientists on the ground knew exactly what was happening to the instrument payload at all heights.

By 1953, then, with powerful rocket boosters in the process of development, it began to seem that some object, even if perhaps only a small one, could be put into orbit.

At the time an International Geophysical Year was being planned for the period from July 1, 1957 to December 31, 1958, one in which there was to be a large international effort to study the properties of the terrestrial globe as a whole. There was a feeling in the United States that it would be appropriate to place a satellite into space in the course of that period. After all, how better to study Earth as a whole than from an object orbiting it?

The Soviet Union announced a satellite program of its own at the time, but the United States did not attach much importance to that, for there was a settled American notion not only that the Americans were the technological leaders of the world (which was true enough) but that the Russian people were particularly retarded in technology and perhaps even innately so (which was not true at all).

The Soviets were aware of this American snobbishness and labored mightily to show it up. Besides, the Soviets were very conscious of the pioneering work of Tsiolkovsky and very proud of it. The centenary of Tsiolkovsky's birth fell on September 17, 1957, and the Soviets labored to place an object in orbit as close to that day as possible.

The Soviets worked quietly, and the Americans cooperated by paying no attention. As a result, the American public and even American scientists and government officials were caught completely by surprise when the Soviet Union, on October 4, 1957 (sixteen days after the centenary goal), placed an object in orbit, thus launching the first artificial satellite in history.

The elliptical orbit of *Sputnik 1,* as it was called, brought its height as low as 230 kilometers at its closest point to Earth's surface ("perigee") and up to 940 kilometers at its highest point above the surface ("apogee"). It weighed 84 kilograms.

# 9
# Out into Space

## THE EARLY SATELLITES

In a broad sense it didn't matter which nation launched the first satellite, any more than it mattered which nation first discovered the American continents. What counts is that human beings achieved an expansion of the human horizon. Any "lead" achieved by this nation or that because of its head start could be overcome and evened out in the long run. And, in the long run, the whole world could benefit, whoever happened to be first.

Just the same, the United States was humiliated at having the Soviet Union first in the field, and more than a little frightened, too, since many Americans were naive enough to believe that somehow to launch a satellite was to threaten to dominate the world.

That had its useful side, however. The American Congress, which might not have voted a penny merely to increase human knowledge, gladly voted billions to "beat the Russians." And the American public, with that as the motive, did not object, so that advances in space followed at a far greater pace than they might have if the Soviet Union had been inactive.

The Russian word *Sputnik,* used for the satellite, briefly became the most famous word in the world, and for a while even Americans used it as a generic term for an artificial satellite. It was widely translated in the United States as "fellow-traveler," which was a phrase that had unpleasant connotations in the anti-Communist lexicon. However, the original Latin meaning of "satellite" is a person who tags along after someone more powerful and who is, therefore, literally, a fellow-traveler. In short, *sputnik* is merely Russian for "satellite."

*Sputnik 1* showed it could be done. On November 3, 1957, a month after the first launching, the Soviet Union showed it could be done again. *Sputnik 2*, a larger satellite, weighing 510 kilograms, was placed in orbit. It was large enough to carry a living dog, so it was the first life-bearing satellite. It showed that a fairly large mammal could survive the accelerations required in the launching process.

Meanwhile the American satellite program was going into high gear, and on January 31, 1958, the first American satellite, *Explorer 1*, was launched. The fact that it weighed only 14 kilograms showed the superiority of Soviet rocket boosters at the time.

Nevertheless, small as it was, *Explorer 1* made an important scientific discovery. It had a counter aboard that could detect high-energy electrically charged particles. It detected such particles in considerable numbers, as was to be expected. After all, surface studies of such phenomena as the aurorae had showed that the particles were there.

*Explorer 1*'s elliptical orbit brought it close to Earth's surface at one end and raised it to a height of 2,500 kilometers at the other end. At higher altitudes, the particle count fell off and, indeed, dropped to zero.

The same observation was made by *Explorer 3*, which was launched on March 26, 1958, and which reached a maximum height of 3,400 kilometers, and by the Soviet satellite *Sputnik 3*, launched on May 15, 1958.

The American physicist James Alfred Van Allen (1914–    ), who was in charge of this portion of the satellite experimentation, suspected that the fall-off was not a real phenomenon. He felt that the reverse was true: that the density of high-energy charged particles increased with height and reached a level that overwhelmed the counters, leaving them unable to function. It was as though the human eye, blinded by light intensity too great to register, could see nothing at all.

When *Explorer 4* was launched on July 26, 1958, it carried special counters designed to handle heavy loads. One of them, for instance, was shielded with a thin layer of lead (analogous to dark sunglasses serving to protect the eyes) that was intended to keep out most of the radiation. This showed that the "too-much-radiation" theory was correct. *Explorer 4* reached a height of 2,200 kilometers and registered an intensity of high-energy radiation far higher than scientists had expected.

It turned out that Earth is surrounded by belts of high-energy radiation consisting of particles originating from the sun and trapped in the lines of force of earth's magnetic field. Originally, these were termed "Van Allen radiation belts," but eventually they were termed the "magnetosphere."

Soon after *Explorer 1* was launched, the United States launched its second satellite, *Vanguard 1*, on March 17, 1958. It was merely 1.4 kilograms in weight, and there were some sneers to the effect that the United States had launched an orange. However, size alone is no measure of value. By carefully studying its orbit and noting minor deviations from the ideal, the American physicist John Aloysius O'Keefe (1916–    ) could detect slight variations in gravitational intensity over various portions of Earth. This showed that Earth itself deviated in minor ways from the ideal oblate spheroid it should have been. The use of *Vanguard 1* and later satellites made it possible for scientists to deduce the exact shape of Earth to within a meter or less in a way that would have been difficult, or actually impossible, from surface measurements alone.

*Vanguard 1* required a source of power to keep signaling surface stations on Earth, if its orbit was to be precisely observed. This power took the form of "solar cells," thin layers of material which on exposure to sunlight developed small electric currents sufficient to activate the radio beam. It was the first use of solar cells in space.

New developments came quickly. *Sputnik 3* weighed 1,330 kilograms (1.33 tonnes) and was large enough to be the first satellite to carry a variety of instruments so that observations of many kinds could be made.

Nor was it long before humanity as a whole became aware of the uses of the new expansion of the human horizon. On December 18, 1958, the United States launched *Score,* which could receive and transmit signals that carried the human voice.

*Score* was merely the beginning. On August 12, 1960, the United States launched *Echo 1,* which expanded into a large aluminum balloon that served as a relay for signals that carried not only sound but pictures as well. Then on July 10, 1962, the United States launched *Telstar 1,* which was not merely a relay but an amplifier as well.

*Telstar 1* was the first true "communications satellite," and in the years since, these have multiplied in number, capacity, and versatility until television coverage is now global. Thanks to communications satellites, Earth is now a unit and no portion is more than a fraction of a second away from any other if human beings trouble to set up the appropriate signaling and receiving stations.

What is more, from the vantage point of space, human beings

were able, for the first time, to look at Earth as though it were a planet like any other; to see it from the outside, so to speak. The United States, on February 17, 1959, launched *Vanguard 2*, which was the first that had the capacity to transmit photographs of Earth's cloud cover that could be received by television. *Explorer 6*, launched by the United States on August 7, 1959, was the first to send back photographs of Earth as a whole.

On April 13, 1960, the United States launched *Transit 1B*, the first satellite equipped to take repeated photographs of Earth and send them back. It took nearly 23,000 photographs in its lifetime, and scientists on Earth had a steady view of the changing cloud cover and could study the atmospheric circulation. It was the first "weather satellite."

Until then, weather forecasting had been a dreadfully incomplete art based on what few reports could be obtained from the small fraction of Earth's surface that was equipped to report local weather conditions in detail. These were missing from all the oceans and from vast sections of the continents.

From 1960 on, however, Earth as a whole was continually under observation, and the art of weather forecasting advanced considerably. As an example, prior to 1960, there was no way of knowing when hurricanes were forming, where exactly they were located, or where precisely they were heading, except for reports from occasional ships that had been caught in one. After 1960, the birth, development, and course of all such storms everywhere were known and, as a result, untold numbers of lives were saved and enormous property damage prevented. (This is one of the many practical advantages brought to humanity by the space program, though it is fashionable on the part of many to deride the program as merely a waste of money.)

On April 1, 1960, the United States launched *Tiros 1*, a satellite intended to serve as a navigation aid. Others of the sort followed. The signals they emit can be received and from the known position of the satellite at any given time, a ship can locate itself on Earth's surface without having to sight any star, hence without having to depend on clear skies. A ship, properly equipped, will know its location, at all times, within meters. It also makes it possible to map Earth with an accuracy formerly impossible.

Numerous satellites were also sent into orbit to serve military purposes, to detect nuclear explosions or mass movements. Although this seems to be a warlike endeavor, it can be viewed as serving the cause of peace. The spy satellites make it more difficult than otherwise for either the Soviet Union or the United States to prepare a preemptive nuclear strike without the other becoming aware of what was going on. This decreases the likelihood of such a strike.

Unfortunately, there is increasing discussion of placing sophisticated weapons in space. This might offer one side or the other a short-term military advantage, but it would merely increase the probability (already far too high for comfort) of suicide for civilization as a whole in the long run.


## PEOPLE IN ORBIT

Though enormous feats can be performed by totally inanimate satellites carrying complex miniaturized equipment for observation and telemetering, such an extension of the observational horizon inevitably brings up the question of whether people can follow.

*Sputnik 2,* with its dog on board, had shown that living things could be carried into orbit without harm. There had been, however, no way of retrieving the dog, and, in time, it had been deliberately poisoned to prevent suffering.

It was not till August 19, 1960, that the Soviets did better in this respect. They launched *Sputnik 5,* which weighed 4.6 tonnes and was the most massive satellite yet put into space. It carried two dogs and six mice, all of which were eventually successfully retrieved.

After that it was only a matter of time before a human being was put into orbit and safely brought back. On April 12, 1961, the Soviet Union launched *Vostok 1,* which had a mass of 4.7 tonnes and which carried a human being, Yuri Alekseyevich Gagarin (1934–68). He became the first person to travel through space. (Americans would have called him an "astronaut" but the Soviets use the term "cosmonaut" instead.)

Gagarin was carried once about Earth and was successfully retrieved. It marked a formidable milestone indeed. The first circumnavigators of Earth had been the eighteen men of Magellan's crew who survived to complete the journey in 1522, 1,084 days after they had first put to sea. Now, four and a third centuries later, a human being had completed the circumnavigation of Earth in 108 minutes.

In the time it had taken those eighteen gaunt survivors to sail once around Earth, Gagarin, assuming he could have been kept alive and in orbit, would have been able to go around Earth 14,453 times.

What's more, Gagarin's feat brought that stage of human endeavor to a close. Circumnavigating Earth in a satellite in low orbit is the fastest way one can reasonably move about the planet, and there is not much point in seeking to improve on Gagarin's mark.

The Soviet Union proved that Gagarin's flight was no accident when, on August 6, 1961, they launched *Vostok 2* with Gherman Titov (1935–    ) aboard. He remained in space for no less than seventeen orbits over a period of 25 hours and 11 minutes, a little more than a full day. Then he, too, was successfully recovered, none the worse for his experience.

Quite clearly, human beings could be launched, placed in orbit, kept there for appreciable periods of time, then be safely recovered.

The United States, after sending up two astronauts in suborbital flights and recovering them safely, placed the first American in orbit on February 20, 1962, when *Friendship 7* was launched.* The astronaut on board was John H. Glenn, Jr. (1921–    ). He completed three orbits, remaining in space for 4 hours and 55 minutes. On May 24, 1962, that, too, was duplicated when *Aurora 7* was launched and carried M. Scott Carpenter (1925–    ) through three orbits.

Both nations carried through additional orbital flights. On June 16, 1963, the Soviets launched *Vostok 6,* which carried Valentina V. Tereshkova (1937–    ), the first woman (and, as of this writing, still the only woman) to be placed in space. She remained in space through forty-eight orbits, lasting nearly three days.

Astronauts in orbit about Earth are in free fall and experience what is, in essence, zero gravity. This is not something that is totally alien to terrestrial life. For small creatures, the size of the smallest mammals or less, gravity has so little effect that it may almost be ignored. For sea creatures, the buoyancy of water largely nullifies many of the effects of gravity.

Human beings, however, are land animals and are large enough for gravity to be an overriding fact. We must avoid falling, not only from a height, but even over our own feet while walking. More than that, gravity affects our physiology. Our muscle tone is surely affected by the need for many of our muscles to work constantly to keep us in balance. The muscle action in turn affects bone formation. The circulation of the blood is designed to take gravity into account when we stand in the normal way, so that we become quite uncomfortable if we remain standing on our heads for any length of time.

There was no way of telling in advance just how an extended period of zero gravity might affect the physiology and biochemistry of the body.

Certainly the initial orbital flights demonstrated that zero gravity over a period of a few hours, or even a few days, was satisfactorily

---

* The earliest vessels carrying astronauts in the American "Mercury program" were all given a 7 numeral because the first American astronauts were seven in number.

borne by the human body, but if ever human beings were to penetrate space for anything more than as a kind of circus-trick performance, we would have to know the effects of longer exposure. The
longer human beings can survive zero gravity, the more likely it will
be that someday work in space will become a reality.

For a period of time, spaceflights became endurance contests for
that reason. On June 14, 1963, Valery F. Bykovsky (1934–     ),
piloting *Vostok 5,* orbited Earth eighty-one times and remained in
space under continuous zero gravity for almost exactly five days.

The record was surpassed when the United States, on August 21,
1965, launched *Gemini 5,* which orbited Earth 128 times and remained in space almost eight days. This record was in turn broken,
over and over, and at the time of writing, the record is held by three
Soviet cosmonauts who have remained in space for just about six
months.

Physiological changes did take place, the most disturbing of which
was the gradual loss of calcium from the bones. These changes, however, did not prevent people in space from working effectively. Nor
did they prove irreversible, for astronauts seemed in no way inca-
pacitated after their return, or unable to live a normal life.

Valentina Tereshkova, the one woman cosmonaut, married a male
cosmonaut, Andriyan G. Nikolayev (1929–     ), who had been
launched into space in *Vostok 3* on August 11, 1962. He had spent
four days in space and she three days, but they had a perfectly
normal baby.

To be sure, many normal human situations have not taken place
at zero gravity. Notably, a woman has not engaged in sex, been
impregnated, remained pregnant to term, and given birth to a baby,
all while at zero gravity. As long as human beings are confined to
orbiting Earth in near space, however, it is not vital to determine
whether the full range of human experience can take place at zero
gravity. We can imagine people living and working in shifts in space,
returning always to Earth as the true habitat.

This would, of course, be the simplest of the possible scenarios,
but it is surely a gain that space exploration has shown that at least
the simplest of the scenarios is possible. Had zero gravity proved to
be quickly fatal, or even quickly incapacitating, the expansion of the
human horizon into space would have become enormously more
difficult. Indeed, all thought of such expansion might have had to be
abandoned.

If space exploration was to become practical, spaceships would
clearly have to become reasonably roomy. In the first couple of years

of the people-in-space program, the vessels involved had barely room for one person.

On October 12, 1964, the first spaceship carrying more than one person was launched. It was the Soviet *Voshkod 1,* an advanced version of the *Vostok* and, by a small margin, more massive than the later Vostoks. It carried a crew of three men.

The first American spaceship carrying more than one person was launched on March 23, 1965. It was the *Gemini 3* and carried Virgil I. Grissom (1926-67) and John W. Young (1930-    ). On October 11, 1968, the first American spaceship carrying three people was launched.

To this date no spaceship has been launched carrying more than three people.

Successive spaceships became more maneuverable. They were more than missiles in free fall. They could use small rocket firings in orbit to change position, rising higher or dropping lower, moving forward or lagging behind. One spaceship could rendezvous with another and dock, so that the crew of one could enter the other ship.

On March 18, 1965, the Soviet Union launched *Voshkod 2* with two cosmonauts aboard. One of them was Aleksey A. Leonov (1934-    ). The ship orbited Earth seventeen times, and in the course of the second orbit, Leonov left the ship through an airlock. He was wearing a spacesuit and was safely tethered to the ship. This was the first "space walk," the first time any human being was in space with no greater protection than a spacesuit. Leonov remained in space for about ten minutes, maneuvering in free fall and taking motion pictures.

On June 3, 1965, the United States launched *Gemini 4* with James N. McDivitt (1929-    ) and Edward H. White (1930-67) aboard. During its third and last orbit, White left the ship and remained in space, suited and tethered, for twenty minutes. He was able to maneuver in space with a small rocket unit, the first person ever to do so.

The significance of these space walks was that they proved that it might well be possible for human beings to engage in repair work, construction work, or other useful labors in space, and do so outside a ship and with no more protection than a spacesuit.

By the early 1970s, France, Japan, China, and Great Britain had placed satellites in orbit, but their space achievements remained far behind those of the United States and the Soviet Union.

Space exploration was not carried through without loss of life In June of 1971, three Soviet cosmonauts in *Soyuz 11* were completing a till then wholly successful flight when a minor misfunction of a door seal as they were reentering Earth's atmosphere allowed the atmosphere within the satellite to escape. When the satellite was recovered, the three were dead.

There were no deaths in flight for the Americans, but on January 27, 1967, during a rehearsal of a satellite launch on the ground, a fire broke out within the capsule. It went out of control almost at once in the pure-oxygen atmosphere, and killed three American astronauts. The dead were Virgil Grissom, who, two years earlier, had been captain of *Gemini 3,* the first American satellite to carry more than one person; Edward H.White, who two years before in *Gemini 4* had been the first American to space-walk; and Roger B. Chaffee (1935–67), who had not yet had a chance to be in space.

## THE LUNAR PROBES

But would the human horizon be confined to orbits about Earth in the upper atmosphere?

Not at all. Almost from the start, efforts were made to launch rockets that attained not only the orbital velocity of 8 kilometers per second, but the escape velocity of 11 1/4 kilometers per second.

In 1958, the United States sent three satellites into space, attempting with each one to exceed the escape velocity. All fell short.

The Soviet Union, on January 2, 1959, succeeded in exceeding the escape velocity with *Luna 1.* It was the first satellite that did not take up an orbit about Earth. It receded indefinitely, and radio contact was maintained with it for 597,000 kilometers, or 1 1/2 times the distance of the moon.

It did not, however, exceed the escape velocity from the distant sun, so it did not move out of the solar system altogether. It took up an orbit about the sun as the first "artificial planet."

Rockets that are launched and do not go into orbit about Earth are clearly not satellites. They are called "probes." *Luna 1* was aimed in the direction of the moon, so it was a "lunar probe." It was not aimed with perfect precision, however, and passed the moon at a distance of 6,000 kilometers from its surface, a distance only 1 3/4 times the moon's diameter.

The United States launched *Pioneer 4,* its first rocket to exceed escape velocity, in March 1959. It passed the moon at a distance of 60,000 kilometers.

The Soviet Union managed to correct its aim on September 12, 1959, when *Luna 2* was launched in the direction of the moon. It smashed right into our natural satellite—human beings had finally (in a way) torpedoed the moon. For the first time in history, a human-made object existed (although undoubtedly badly smashed) on the surface of a world other than Earth.

Missing the moon did not accomplish much, and hitting it in a "hard landing" accomplished even less, aside from merely demonstrating accuracy of aim. The next step, however, came to more.

The moon turns about Earth, relative to the stars, in 27 1/3 days. It also rotates about its axis in 27 1/3 days. This is not a mere coincidence. The tidal effect of Earth enforces this upon the moon, and the result is that the moon faces only one side toward us at all times.

Because the moon moves about Earth in an ellipse, rather than in a perfect circle, its orbital speed is not constant. The moon's distance from Earth varies a bit as it moves in its elliptical orbit, and it moves a bit faster when it is closer to Earth than when it is farther.

The rotational speed remains perfectly constant, however. This means that the rotation overtakes the orbital motion for half the orbit, then falls behind for the other half. The moon, therefore, seems to wag slightly back and forth like the beams of an old-fashioned scale. This motion is called "libration" from a Latin word for "scale."

As a result of libration we can see almost 60 percent of the moon's surface from Earth (some of it always at the edge and, therefore, badly foreshortened). A little over 40 percent, on the other hand, is forever hidden, and has been so, probably, for billions of years, certainly for much longer than any human or near-human eyes existed to observe the moon.

A lunar probe could be launched at a speed so close to escape velocity that by the time it reached the vicinity of the moon, Earth's gravitational pull would have slowed it to a crawl. The probe could then fall under the influence of the lunar gravity to the extent of moving into orbit about the moon. It would be "captured" by the moon. (This could take place more easily if the probe had the ability to shoot off a small rocket blast at just the right time and in the right direction to alter its course and speed in such a way as to make the capture more likely.)

At the very least, if the orbital corrections could not be made, the probe could loop around the moon once and then shoot off on another leg of its long journey. Even the single loop would carry it around the far side of the moon, and if it could take photographs that could be telemetered back to Earth, then human eyes would see, *for the first time,* objects less than 400,000 kilometers away that had never been seen before. though objects ten million billion times as far away *have* been seen.

This the Soviet Union managed to do with *Luna 3*, launched on October 4. 1959. On the second anniversary of the launching of the first satellite. the state of the art had advanced to the point where a lunar probe looped around the moon and sent back the first photographs. ever. of the far side.

The quality of the photographs was poor and they were few in number. but they were sufficient to dispel certain romantic notions that had existed concerning the far side. It had always been possible to dream that for some reason or other. the hidden side was less forbidding than the side we saw. that it had a bit of atmosphere and water and. perhaps. life.

Not so! It was clear from those first photographs that the surface of the far side of the moon was as desolate as that of the visible side. American photographs of the far side. when they came. confirmed this.

On July 28. 1964. the United States launched *Ranger 7*, which hit the moon as *Luna 2* had done nearly five years earlier. Now, a new refinement had been added. During the last thirteen minutes before impact. photographs were taken and telemetered back to Earth. Altogether. 4.308 photographs were taken at closer and closer range until contact was made and all transmission ceased. For the first time. the moon was photographed at very close quarters, with detail far beyond anything imaginable with an Earth-based telescope (though the portion of the moon that was so photographed was very small indeed).

The photographs showed the moon peppered with craters of very small size as well as of very large size. Some were only a meter wide and 30 centimeters deep.

The United States repeated this with *Ranger 8* and *Ranger 9* in 1965. the two together sending back 13,000 photographs of two other places on the moon.

Through 1965. those probes that had landed on the moon had done so by way of a "hard landing"—one that resulted in the destruction of the object that came to rest there. What was needed was a "soft landing," one that would leave the probe still functioning on the moon's surface.

If the moon had had an atmosphere that task would have been relatively simple. since a parachute could have been deployed to help slow the falling probe. Without an atmosphere, the slowing would have to be done entirely by rocket blasts firing in the direction of the moon's surface and thus pushing the probe away from the moon and slowing it down. That would require the most delicate handling.

The Soviet Union achieved the task on February 3, 1966, when its probe *Luna 9* made a soft landing on the moon. A little over six years after the first hard landing on the moon, a human-made object was resting on the moon's surface sufficiently intact to allow its instruments to work. The first photographs ever to be taken on the moon's surface were received on Earth.

*Luna 9* had landed in the western part of Oceanus Procellarum, a relatively uncratered portion of the moon's surface. The name of the region is Latin and means "Ocean of Storms"—an indication of the fanciful notions of the early astronomers, before it was quite realized that the moon was waterless and airless. The pictures, which were sent to Earth for three days before the instruments failed, showed neither ocean nor storms (of course) but merely a stony region of quiet desolation.

The United States launched *Surveyor 1* on May 30, 1966, and it duplicated the Soviet feat. It made a soft landing on the moon in Oceanus Procellarum also, and sent back photographs of somewhat better quality than those of the Soviet probe.

Later American satellites went further. *Surveyor 3,* launched on April 17, 1967, not only made a soft landing and returned more than 6,300 pictures of the lunar surface, but used a sampler arm to dig in the lunar soil in automatic response to signals from Earth.

This was important, since there had been some suggestions at this time that the moon was covered with a thick layer of loose dust produced by the fragmentation of the surface through meteoric bombardment over the ages, and that any vessels landing on the moon would sink into that dust layer and disappear, rather as though it were dry quicksand. The mere fact that soft-landed probes remained on the surface argued against that, and the digging of the sampler arm clearly showed that the lunar soil possessed the same order of consistency as earthly soil.

*Surveyor 5,* which landed on the moon in Mare Tranquillitatis ("Quiet Sea," which is an accurate description), even carried a small device that would analyze samples of the lunar soil. Altogether the *Surveyor* probes sent back over 50,000 photographs, and the lunar surface, as seen from ground level, became a familiar sight to human beings.

On March 31, 1966, the Soviet Union launched *Luna 10,* which moved into orbit about the moon. The United States duplicated this when it launched *Lunar Orbiter 1* on August 10, 1966. These and other probes orbiting the moon sent back photographs of all parts of the moon in fine detail, with the result that we have maps of the entire moon as reliable and almost as detailed as maps of Earth.

What emerged from these maps confirmed the suspicions that arose with the very first pictures of the far side. Whereas the side facing Earth has a number of "maria," large, roughly circular regions that are relatively free of craters, with floors that seem to have arisen as lava flows in the dim early ages of the moon's history, the far side is almost free of such things.

The moon's crust is roughly 60 kilometers thick on the near side, but about 100 kilometers thick on the far side, as nearly as we can tell from what we have learned about the moon through our probes. As a result, a large meteor strike on the near side could collapse the crust and allow heated rock to bubble up from the molten layers beneath, whereas a similar meteor strike on the far side would be withstood by the thicker and stronger crust.

But why should the crust be asymmetric in this fashion? It seems natural to suppose that Earth had something to do with it, but no convincing suggestions have yet been made as to what that something might be.

One discovery that was not astonishing was that the moon does not have a magnetic field. Current theories for the existence of magnetic fields about astronomical bodies suggest that there are two requirements. First there must be, at the center of the body, a liquid core capable of conducting an electric current. Second, the body must itself be rotating rapidly enough to set up swirls in the liquid at the core. The circular movement of electric charge would then, in turn, set up a magnetic field.

Earth has a liquid core rich in iron, and it rotates rapidly, so it is not surprising that it has a magnetic field.

The moon, on the other hand, has a density only three-fifths that of Earth. It must be far less rich, all told, in relatively dense substances such as iron. The moon, moreover, being considerably smaller in mass than Earth, is sure to be less hot at the center. (The central heat arises from the motion of the smaller bodies that crashed together to form a large body, and the moon's relatively small gravity could only produce slow speeds which would be converted to relatively little heat.)

The combination of relative lightness and coolness makes it reasonably certain that the moon lacks a liquid iron core and is probably solid rock through and through. Even if there were a liquid iron core capable of conducting electricity, the moon's rotation is far too slow to set up significant swirls in that core. All told, then, if our theories are correct, the moon ought to have no magnetic field—and it has none.

## REACHING THE MOON

But if unmanned probes skimmed the moon, circled it, landed on it—could human beings follow?

The United States early announced that as a goal. It had been stung by having been steadily forestalled and overshadowed by the Soviet Union in what came to be called the "space race." In addition, the United States had suffered a humiliation on April 17, 1961, when an attempted invasion of Cuba by American-supported anti-government Cubans at the Bay of Pigs was smashed.

To retrieve American prestige, President John F. Kennedy (1917–63) announced, in May 1961, that the United States would place a human being on the moon before the end of the decade. With that end in view, the "Apollo program" was established by the National Aeronautics and Space Administration (NASA).

The "Gemini program," which preceded it, with its emphasis on space maneuvering, rendezvous, dockings, and space walks, was intended to develop the sort of capabilities that would be required for the moon landing. The Gemini program was completed by the end of 1966 and the first Apollo flight was scheduled for February 1967.

It was delayed by the tragic death of the three astronauts Grissom, White, and Chaffee, who were engaged in work in the Apollo spacecraft on the ground. The spacecraft had to be redesigned. Several unmanned Apollo launches were then successfully carried through.

Finally, the first manned Apollo craft (with three astronauts aboard) was launched on October 11, 1968. It was *Apollo 7* under the command of Walter M. Schirra (1923–    ). It was the United States' first three-man spaceship, and Schirra's third venture into space. In the course of 163 orbits, the guidance and control systems of the Apollo design were tested, as was the ability of the rocket engines to restart on command. Only a little over a year was left now, if President Kennedy's promise was to be fulfilled.

Two months later, on December 21, 1968, *Apollo 8* was launched, under the command of Frank Borman (1928–    ). Borman at that time held the record for space endurance, for in *Gemini 7*, three years earlier, he and James A. Lovell (1928–    ) had remained in orbit for 220 revolutions, or just about two weeks. Another feature of that trip had been that *Gemini 6,* launched eleven days after *Gemini 7,* had, under Schirra's command, successfully rendezvoused and docked with *Gemini 7,* a maneuver that would be necessary in a lunar landing.

Now *Apollo 8* attempted an unprecedented maneuver. The three astronauts took the vessel to the moon and placed it in orbit about that body. For the first time in history, human beings were close enough to the moon to study it at leisure with the unaided eye, seeing detail no telescope based on Earth's surface could show. *Apollo 8* remained in the neighborhood of the moon for ten orbits, or a little over six days. It then returned safely to Earth, having remained in space altogether for a little over ten days. It was the most spectacular space feat up to that time.

*Apollo 7* and *Apollo 8* had remained intact during their entire stay in space, though each actually consisted of two spaceships. It was intended, eventually, after going into lunar orbit, to have one ship, the "lunar module," with two men aboard, separate from the other, the "command module." The lunar module would then move down to the surface of the moon, while the command module remained in orbit. Eventually, the lunar module would leave the moon's surface and dock with the command module again.

The astronauts in the lunar module would then enter the command module. The lunar module would be discarded and the command module, with all three astronauts on board, would then return to Earth.

On March 3, 1969, *Apollo 9* was launched under the command of McDivitt and the lunar module was checked out in space for the first time, but for caution's sake, *Apollo 9* remained in Earth orbit throughout. It was in space for 151 orbits or nearly ten days.

The next step was taken on May 18, 1969, when *Apollo 10* was launched, under the command of Thomas P. Stafford (1930–    ), who had been in space twice before and had been with Schirra on the occasion of the first successful docking.

*Apollo 10* repeated the feat of *Apollo 8,* placing itself in lunar orbit on May 21 and remaining there for thirty-one orbits. This time, the lunar module separated and carried Stafford, along with Eugene A. Cernan (1934–    ), down toward the surface of the moon. (Cernan had been with Stafford on the flight of the *Gemini 9* three years before, and on that occasion Cernan had maneuvered in a space walk for over two hours.)

Stafford and Cernan swooped down to within 15 kilometers of the moon's surface, then returned to the command module, in which John W. Young (1930–    ), who had been on two of the Gemini flights, waited. The spaceship returned safely to Earth after eight days in space.

All had now been done except for the actual landing.

On July 16, 1969, *Apollo 11* was launched with three astronauts aboard. These were Neil Armstrong, Edwin E. Aldrin, and Michael Collins. All three had been born in 1930.

Armstrong had been captain of *Gemini 8*, which had been launched on March 16, 1966. He had docked with an unmanned rocket, but a malfunction did not allow the maneuver to be carried through to the planned completion.

Aldrin was on *Gemini 12*, the last of the series, launched on November 11, 1966. In the course of that four-day flight, Aldrin undertook a space walk lasting five and a half hours.

Collins was on *Gemini 10*, launched on July 18, 1966, which had docked with an unmanned rocket more successfully than *Gemini 8*, but still not with full perfection.

Now each of the three were on their second mission into space. On July 20, 1969, when *Apollo 11* was moving smoothly about the moon in an orbit that kept it at a height of 100 to 120 kilometers above the lunar surface, the lunar module, with Armstrong and Aldrin aboard, sank downward to the surface near the southwestern edge of Mare Tranquillitatis.

At 4:18 P.M. Eastern Daylight Time, on July 20, 1969, the lunar module, which was under Armstrong's manual control, made a safe landing on the moon. For the first time in history, living beings from Earth were on the surface of another world. The landing on the moon had been made in the decade of the 1960s, as President Kennedy had predicted, with 164 days to spare. Kennedy himself, however, did not live to see the triumph. He had been assassinated on November 22, 1963. It was Richard M. Nixon (1913–    ), the man Kennedy had defeated for the presidency, who now sat in the White House.

Armstrong emerged from the lunar module and was the first human being actually to set foot on the moon, saying, as he did so, "That's one small step for a man, one giant leap for mankind." Aldrin followed him. The whole procedure was viewed by hundreds of millions of breathless human beings, glued to television sets all over Earth.

The lunar soil supported them and the module without trouble. They spent a total of 21 hours and 36 minutes on the moon, taking photographs, collecting samples of rock, planting an American flag, and setting up experiments that would telemeter results back to Earth after they were gone.

They then lifted the lunar module off the surface of the moon and returned to the command module, where Michael Collins had been continuing to orbit the moon. All three astronauts were safely back on Earth on July 24.

Six more trips to the moon were made in the course of the next three and a half years. The most dramatic was that of *Apollo 13*, launched on April 11, 1970, under the captaincy of Lovell, who had been on *Apollo 8* and was now on his fourth space mission.

It was the only Apollo flight aimed at the moon that did not make it. An explosion in a tank of oxygen resulted in the loss of all power in the service module, which was below the command module and which contained the propulsion system for midcourse corrections, for achieving lunar orbit, and for returning from the moon. It meant there was no way of landing on the moon. Using oxygen and power in the lunar module, the astronauts managed to swing about the moon and return safely to Earth.

*Apollo 15*, launched on July 26, 1971, under the command of David R. Scott (1932–    ), who had been on *Gemini 8* and *Apollo 9*, landed on the moon on July 29. It carried a "lunar rover," a vehicle designed for traveling on the moon. Two years after human beings had first set foot on the moon, human beings drove what was, in essence, an automobile about the surface of the moon. They drove the rover three times for a total distance of 28 kilometers.

In the six actual landings on the moon, hundreds of kilograms of moon rocks were collected and brought back to Earth.

There have been no moon landings since December 1972, and none are being planned at the moment. Nevertheless, it is plain that the human horizon has been expanded to include the moon. Human beings can surely return to the moon anytime they wish to expend the effort and the resources to do so.

# 10
# The Inner
# Solar System

## MAKING USE OF SPACE

If humanity is to return to the moon, it is certain not to be under the circumstances of the Apollo program. That program used Earth itself as a base, and it also used primitive spaceships that were discarded after one use. That was good enough for the first bits of exploration, but there must be better for the next stage.

In order to do more, it must be determined just how well human beings can function in space over prolonged periods, reusable spaceships must be devised, permanent bases in space must be set up, and an elaborate space technology must be worked out.

The United States launched its first space station (a satellite designed to remain in space for years and to house astronauts for extended periods) on May 14, 1973. This space station, named *Skylab*, was more or less cylindrical, was 15 meters long and 6.6 meters wide, and was divided into two rooms. It had a mass of 70 tonnes.

*Skylab* was occupied on three separate occasions, each time by a group of three astronauts. The first group remained in *Skylab* for twenty-eight days, the second for fifty-nine days, and the third for eighty-four days. The final mission was completed on February 8, 1974. A number of scientific experiments were carried out, but the most important of all was the mere fact that for a month, for two months, and then for three months, human beings remained in space under zero-gravity conditions, functioned normally, and incurred no major physical or mental problems.

After the third mission, *Skylab* continued to circle Earth, untenanted and unused, for five more years. It moved through the exosphere at an original height of 425 kilometers or so, where there were present enough wisps of upper atmosphere to sap its kinetic energy and bring it down eventually.

This was known at the time it was launched. It was estimated that *Skylab* would remain in orbit for ten years and that by that time it would be possible to send up special ships designed to lift it higher.

Unfortunately, the ship that might have been used was delayed in its development, while the sun, more active at sunspot maximum than usual, heated and expanded the upper atmosphere a bit more than normal, subjecting *Skylab* to a greater-than-expected air-resistance drag.

By late 1978, it was clear that *Skylab* would plunge to Earth sometime in 1979, and that there was no way of stopping it. Satellites and their fragments had come to Earth before; a Soviet satellite, carrying nuclear components, had crashed in a remote area in Canada some years earlier, for instance. However, *Skylab* was by far the largest human-made object to drop out of space, and there was widespread fear that it might do damage. Assurances by NASA that there was a vanishingly small chance of this, and that the probability of being struck by a fragment of *Skylab* was roughly equivalent to the chance of being struck by a meteorite, did not noticeably calm the prevailing nervousness.

Finally, on July 11, 1979, *Skylab,* having completed 34,980 orbits, plunged to Earth on its 34,981st. The scientists in charge did their best to so control matters as to ensure a plunge into the ocean. Most of it did fall into the Indian Ocean, but a number of pieces managed to make it to the deserts of western Australia. No damage was done and there were no casualties.

Meanwhile, from 1975 on, the Soviet Union had space stations of its own. These were used much as the United States used *Skylab,* but over a longer period and even more successfully. On two separate occasions, crews of three remained in space continuously for a period of half a year. One cosmonaut was on each of these missions and therefore remained in space at zero gravity for a year all told. Again, no major physical or mental problems were reported.

There seems no doubt, then, that human beings can live and work for extended periods in a space environment. What must next be done is to improve our spaceships.

For this reason, both the United States and the Soviet Union are developing space shuttles. These are vessels intended to be reusable. They are designed to move into space and return, over and over.

The American space shuttle was plagued by difficulties and by the intense attempts to make certain that all technical problems were solved. For instance the ceramic tiles covering the shuttle (31,000 of them) had to remain firmly bound to the ship under the most stringent conditions in order that they might surely resist the heat and atmospheric drag on the return trip.* After all, this was the one major new venture in manned spaceflight that was not to be preceded by unmanned tests, and the crew must be protected at all costs.

As a result, the shuttle program was repeatedly delayed and the estimated date of the first flight repeatedly put off. Finally, on April 12, 1981, the first operational shuttle, *Columbia,* lifted off perfectly and, two days later, landed perfectly, while the nation watched and roared its approval. Later in the year, a second shuttle flight was safely carried through, though not everything worked perfectly.

Undoubtedly, more shuttle flights will be made and the shuttles will be used for parking satellites in orbit and for bringing structural components into space for later fabrication into large structures that could not practicably be launched in one piece.

In doing this, it seems reasonable to suppose that we will build structures that will help us take advantage of the resources that nearby space will make available to us.

For one thing, there is the solar energy radiating ceaselessly in our direction. The fossil fuels on which we chiefly rely for energy will not last forever and, in particular, oil, a copious energy source, will not long survive the twentieth century. Already, then, there is a strong movement in favor of the development of solar energy as an alternate source.

To collect the energy at the surface of Earth has certain disadvantages. Any collecting station on the surface would be in the night shadow of Earth twelve hours out of the twenty-four, on the average. Even clear air absorbs a sizable fraction of sunlight, and this becomes more accentuated the farther the sun is from zenith and the dustier or mistier the atmosphere.

One can visualize, instead, a solar power station in orbit about Earth. An array of photovoltaic cells, converting radiant energy into electricity, will be exposed to the full range of solar radiation with no atmospheric interference at all. If it is in synchronous orbit (some 35,000 kilometers above Earth's surface) in the equatorial plane, it will turn as Earth does and seem to hover over one spot on Earth's equator. It will then encounter Earth's night shadow only during the period about the equinoxes in March and September. It will be in darkness for a total of only about a week a year.

* The tiles give the shuttle a brick-walled appearance oddly reminiscent of Hale's *The Brick Moon.*

It has been estimated that an array of photovoltaic cells might convert sixty times as much solar energy into electricity as the same array would on Earth's surface. The electricity formed in space could be converted into microwaves and beamed to Earth's surface, where it could be collected much more efficiently than the original sunlight would have been.

It would be expensive and difficult to set up a series of such solar energy stations in space, but once set up they might very well quickly pay back the expense and become an essential part of Earth's technological civilization.

Other useful structures could be built in space as well. There could be astronomical observatories, of course. There could also be research laboratories taking advantage of those properties of space difficult (or impossible) to duplicate on Earth's surface. Such properties include the presence of indefinite volumes of hard vacuum; high and low temperatures, depending on whether an object is exposed to sunlight or is shaded from it; hard radiation, if exposed to sunlight; and gravity-free conditions. Even isolation from Earth by thousands of kilometers of vacuum is a useful property, since dangerous experiments could be performed out there rather than on Earth's crowded surface.

Many industrial laboratories, each highly automated and roboticized, might be located in orbit. In that case, the inevitable waste products and pollution of industry could be discharged into space rather than into Earth's biosphere.

The advantages of this are that space is a far less delicate environment than the biosphere is and is far less likely to be damaged; and that space has enormously more volume than the biosphere has and could far more effectively dilute the wastes. Nor would the wastes even remain in Earth's vicinity, since the solar wind (a stream of high-energy charged particles streaming ceaselessly from the sun in all directions) would sweep them out into the far greater and more distant volume of the outer solar system and beyond.

Thus, the development of space might retain for us all the advantages of a highly industrialized society while removing the industries themselves (to some extent, at least) from our immediate midst and thus reducing the disadvantages.

The metals, concrete, and glass needed to build all these structures in space could well come from the moon, rather than from Earth itself. The only vital elements which the moon lacks are carbon, nitrogen, and hydrogen and these three are in good supply on Earth, which can serve as an interim source until our space technology can spread its wings even farther than the moon.

Finally, to speak of immaterial things, the development of a space technology might well be carried out by the nations of Earth in cooperation, since they will all benefit mightily from the results. This could serve as a positive factor in encouraging still further cooperation, and in putting an end to the warfare that has always engaged human beings and that has now reached a pitch of intensity and power that can certainly wipe out civilization and might even put an end to humanity in general.

This may seem a Pollyannalike view of things, but so dreadful is the world situation today that it would not hurt to try it and see.

## VENUS

All this might yet come to pass, even if Earth's expanding horizon, which has now reached the moon, should expand no farther. But could it possibly extend no farther? Having reached the moon, will we stop? *Need* we stop?

The difficulty lies in distance.

The moon's distance from Earth is anywhere from 356,000 to 407,-000 kilometers, depending on its position in its orbit. This is not really a very great distance. On the average, the distance to the moon is only about 9.5 times Earth's circumference.

This lunar distance was first understood by the ancient Greek astronomer Hipparchus of Nicaea (190–120 B.C.), as long ago as 150 B.C. Greek astronomers, however, lacked the instruments to obtain a good notion of the distance of any other heavenly body.

It was not until the telescope was first used by Galileo in 1609, and an accurate clock was first devised by the Dutch physicist Christiaan Huygens (1629–95) in 1656, that planetary distances could finally be determined. The first reasonable values were obtained by the Italian-French astronomer Giovanni Domenico Cassini (1625–1711), in 1672. It was only then that the vast extent of the solar system was made plain.

Consider, for instance, the other planets of the inner solar system—bodies rather like Earth or the moon in overall chemical structure. There are three of them: Mercury, Venus, and Mars.

Of the three, Venus is our closest neighbor. Its actual distance depends on the particular position of Earth and Venus in their respective orbits. When both are on the same side of the sun and in a direct line with the sun, Earth and Venus are then at their closest to each other. Even then, however, they are separated, at the very least, by 38,900,000 kilometers. At their closest approach to Earth, Mars is

56,000,000 kilometers away and Mercury is 80,000,000 kilometers away.

In other words, even at closest approach, Venus is 95.6 times as far away as the moon is at its farthest; Mars is 137.6 times as far away; and Mercury is 196.6 times as far away. Where a trip to the moon takes only three days, a trip even to Venus would take many months, and at the present level of technology that would not be an easy task.

However, if we are dubious as to our own ability to go to these worlds, at least right now, there is no reason we cannot send our instruments, and, indeed, we have done so. Space probes have traveled past each of these worlds, with gratifying results for astronomers.

The first planet to receive attention from such probes was Venus. It approached us most closely and, in any case, had already yielded surprising information.

Venus is perpetually cloud-covered, so nothing of its surface could ever be seen by optical telescopes, no matter how large and advanced they might be, but ordinary light is not the only type of radiation astronomers could use.

As early as 1931, an American radio engineer, Karl Guthe Jansky (1905-50), had detected radio waves reaching Earth from outer space. At the time, astronomers lacked the necessary equipment to deal with these radio waves.

During World War II, however, the necessities of war led to the rapid development of devices for producing and detecting beams of comparatively short radio waves called "microwaves" (which, despite "micro," meaning "small," are still much longer than light waves). These microwaves can penetrate mist, fog, and clouds that light waves cannot. The microwaves can, however, be reflected by solid bodies, much as light waves can be. The reflected microwaves can be detected after reflection, and from the time lapse between the emission of the microwaves and the detection of the echo, the distance (and, of course, the direction) of the solid body that did the reflecting can be determined.

It is by the emission of microwaves and the detection of the reflections that radar works. It was the British who first developed radar to a pitch of usefulness, and it was through the use of radar that the British managed to be ready for German planes and to defeat them in the Battle of Britain in 1940.

After World War II, astronomers took advantage of the radar expertise to develop the necessary equipment for dealing with radio waves generally, and radio astronomy was born.

Beams of microwaves could, for instance, be sent out to other worlds and the reflections detected. The distance of these worlds could then be determined just as easily as the distance of approaching airplanes could be. Microwaves were reflected from the moon by a Hungarian scientist, Zoltán Lajos Bay (1900–    ), in 1946.

Venus is a much more distant object and, therefore, harder to hit, and, also because of the distance, the reflection is feebler and harder to detect. In 1958, nevertheless, Venus was hit by microwaves and the reflection was detected. This made it possible to determine the distance between Venus and Earth at that moment more accurately than had been possible by any other method, and from that all the other distances in the solar system could be calculated more accurately than before. It turned out that all those distances were actually about 0.5 percent less than had been thought.

It so happens that all bodies give off radiation. If they are hot enough, the radiation is energetic enough and short-wave enough to emerge in the form of light waves, at least in part. Thus if you heat an object, it eventually becomes red-hot. If you heat it further, it becomes white-hot.

Objects too cool to be red-hot give off infrared waves, which are longer than ordinary light waves, and less energetic. Infrared waves cannot be seen by the eye but they can be detected by instruments. The cooler an object, the longer the waves it gives off, on the average. From the pattern of waves given off, the temperature of an object can be determined.

Venus's surface gives off a pattern of long-wave radiation, including considerable quantities of microwaves, and whereas ordinary light could not penetrate the planet's cloud layer, microwaves can penetrate it easily.

In 1956, the pattern of microwave radiation from Venus was analyzed, and it at once appeared, from that pattern, that Venus might be hotter than had been thought—even considerably hotter. The usual thought at the time was that Venus, being closer to the sun, would be warmer than Earth, but that since there was a protective cloud layer which reflected at least three-quarters of the sunlight arriving, it might be not very much warmer. The microwave radiation from Venus seemed to contradict this, however.

Over the next couple of years, both the Soviet Union and the United States developed the capacity to place satellites in orbit about Earth, and both began to plan Venus probes.

The first successful Venus probe was *Mariner 2*, launched by the United States on August 27, 1962. It traveled over a curved path that carried it 290,000,000 kilometers in 109 days. Finally, on December 14, 1962, when it was 58,000,000 kilometers from Earth, in a straight line, it skimmed past Venus at a distance of 35,000 kilometers above its cloud layer.

The instruments on *Mariner 2* were able to measure the micro-wave radiation from Venus in great detail from various spots on its globe. It showed that the surface of Venus was hotter than even the first microwave measurements on Earth had indicated, and all later probes have confirmed that.

The surface, we are now certain, is hellishly hot all over Venus, near the poles as well as at the equator, and on the nightside as well as on the dayside. The surface temperature is something like 475° Celsius (890° Fahrenheit),* which is more than hot enough to melt tin and lead and to boil mercury.

Why is Venus so surprisingly hot? The responsibility seems to lie with its atmosphere.

The Soviet Union sent a series of probes to Venus, two of which, *Venera 9* and *Venera 10,* succeeded in striking the planet itself and descending to the surface by parachute in October 1975. They did not last long, for Venus's surface conditions are extreme. They did have a chance to show, however, that the atmosphere of Venus was about ninety times as dense as Earth's.

The probes took photographs of Venus's surface, the first ever received from that planet. They showed a dry, rocky surface that was surprisingly well lit. The clouds in Venus's atmosphere reflected or absorbed up to 97 percent of the sunlight striking the planet, but the quantity that penetrated was sufficient to make Venus's surface as bright as Earth's would be on a cloudy day.

All this was confirmed after the United States launched *Pioneer Venus* on May 20, 1978. It arrived on December 4, 1978, and sent several daughter probes into Venus's atmosphere. Venus's atmo-sphere, as it turned out, was 96.6 percent carbon dioxide and 3.2 percent nitrogen. Minor components make up the rest.

Carbon dioxide has the property of absorbing long-wave radiation and heating as a result, whereas oxygen and nitrogen do not. Earth's relatively thin atmosphere of oxygen and nitrogen let the sunlight pass through during the day to be absorbed by the ground, which heats as a result. At night, Earth radiates that heat into space in the form of long-wave infrared radiation, which also passes through the oxygen and nitrogen of the atmosphere, cooling Earth as much (on the average) as it has been warmed during the day.

Venus's dense carbon dioxide radiation allows any sunlight that makes its way through the clouds to reach the ground and warm it.

---

* Celsius and Fahrenheit are usually abbreviated "C." and "F." respectively. These two ways of measuring temperature will be discussed in detail later in the book (see page 298).

At night, however, infrared radiation from the ground cannot pass through the carbon dioxide, so that Venus does not cool. Its temperature rises; more and more infrared radiation is produced; and finally enough can make its way through the atmosphere to keep Venus from warming any further. By that time, though, Venus has grown very warm indeed.

This is called the "greenhouse effect," because it is sometimes suggested that the same phenomenon keeps greenhouses warm in cold weather (though there the actual effect seems to be a little different).

Thanks to *Pioneer Venus*, astronomers have been able to map Venus's surface in considerable detail despite the eternally obscuring cloud layer. To do this, it is necessary to use microwave beams that penetrate the clouds on the way in, then bounce off the surface to produce a reflected beam which penetrates the clouds on the way out. The reflected microwave beam yields information about the surface as a reflected light-wave beam would. (The microwaves must be analyzed by instrument, of course, whereas the light waves could be analyzed by eye or camera. Then, too, the microwaves, being longer than light waves, yield fuzzier results—rather as though the surface were being viewed astigmatically.)

Most of Venus's surface seems to be level, and of a sort we associate with continents rather than sea bottoms, so that it would seem that Venus never had an ocean even in its early history, but, at the most, some inland seas. The continental area makes up about five-sixths of the entire surface, with small regions of lowland making up the remaining sixth.

The huge Venusian super-continent has some indication of craters, but not very much. The thick atmosphere may have eroded them away. There are, however, raised portions of the super-continent, two of them being particularly large.

In what, on Earth, would be the Arctic region, there is a large plateau on Venus which is named Ishtar Terra ("Land of Ishtar"; Ishtar is the Babylonian equivalent of the Roman goddess Venus).

Ishtar Terra is about as large as the United States. Its western portion is relatively flat and is about 3.3 kilometers above the ordinary level of Venus. In its eastern region is a mountain range, Maxwell Montes, of which the highest peaks are as much as 11.8 kilometers above the average level outside the plateau. These are 3.7 kilometers higher than Mt. Everest on Earth.

In the equatorial region of Venus there is another and even larger plateau called Aphrodite Terra (Aphrodite is the Greek equivalent of the Roman Venus). It is 9,600 kilometers wide and has some peaks that are 8 kilometers high. From the eastern end of Aphrodite Terra, a group of great canyons extend for some 5,000 kilometers. Some of these split the crust to a depth of 2.9 kilometers below the average level of Venus's surface.

A small region of elevation southwest of Ishtar Terra is Beta Regio, which possesses two mountains, Rhea Mons and Theia Mons, each about 4 kilometers high. These may be volcanoes. Rhea Mons may spread across an area as large as New Mexico and would be far larger than any volcano on Earth.

These plateaus, mountain ranges, volcanoes, and canyons make up only 5 percent of Venus's surface.


## MERCURY

The only sizable object that circles the sun at a distance less than that of Venus is the planet Mercury. It revolves about the sun in eighty-eight days, and until 1962 it was thought to rotate on its axis in eighty-eight days as well. This would mean that it faced one side always toward the sun and one always away from the sun.

In 1962, however, the study of microwaves reflected from Mercury's surface showed that it was rotating, actually, in 58.65 days, or in just two-thirds of its period of revolution, so that every part of Mercury's surface experienced day and night.

Mercury's maximum temperature at the spot where the sun is at the zenith at the time Mercury's orbit brings the planet closest to the sun is 425° C. (800° F.). This is not quite as high as the temperature of Venus, even though Venus is farther from the sun—but then Mercury has no atmosphere and therefore there are no gases to store and conserve heat.

Then, too, Venus's temperature stays fairly close to its maximum at all times everywhere on its surface. Mercury, on the other hand, loses heat (since there is no atmosphere to trap it) as the sun declines, and does so particularly once the sun is below the horizon. By the time the sun has been absent through the long night and is ready to rise, the temperature has sunk to −180° C. (−290° F.).

Despite the fact that Mercury has no atmosphere, it is so far from Earth and is always seen so near the sun that Earth-based telescopes have been able to make out virtually nothing of its surface. The use of probes has changed matters.

On November 3, 1973, *Mariner 10* was launched. It passed by the moon and then, on February 5, 1974, it passed by Venus just 5,800 kilometers above the cloud layer. It then headed for Mercury and became the first Mercury probe when, on March 29, 1974, it passed within 700 kilometers of Mercury's surface.

It then moved about the sun in an orbit which gave it a period of revolution of 176 days, or twice the length of Mercury's year. This brought it back to Mercury every time that planet completed two revolutions. *Mariner 10* passed Mercury on September 21, 1974, a second time, and then on March 16, 1975, a third time. On the third pass, it skimmed within 327 kilometers of its surface. After the third pass, *Mariner 10* had consumed the gas which kept it in a stable position and it was thereafter useless for further study of the planet.

The photographs which *Mariner 10* took of Mercury showed a landscape that looked very much like that of the moon. There were craters everywhere, with the largest one observed about 200 kilometers in diameter.

Mercury is not as rich as the moon in marias. The largest one sighted on Mercury is about 1,400 kilometers across. It is called Caloris ("Heat") because it is just about at the spot on Mercury where the sun is overhead when it is closest to the planet.

Mercury also possesses cliffs that are several hundred kilometers long and about 2.5 kilometers high.

*Mariner 10* has photographed only about three-eighths of the surface of Mercury in the course of its three working passes, but astronomers suspect that the remainder of the planet's surface is much like that which has been seen.

## MARS

The remaining planet of the inner solar system is Mars.

Unlike Mercury and Venus, Mars is farther from the sun than Earth is, and it is far more easily observed than Mercury and Venus can be. Mars can at times be considerably closer to Earth than Mercury ever is, and it can be seen high in the sky at night with its entire visible face in sunlight, whereas Mercury and Venus are always rather near the sun and, when on the near side of their orbits, show only a portion of their face in sunlight.

Mars has an atmosphere, but it is very thin and there are rarely any clouds or mist to obscure the surface. Of the planets of the inner solar system, therefore, it is the only one (other than Earth itself, of course) that was mapped by Earth-based telescopes. The map was largely a pattern of light and shadow (which some thought to represent land and sea). In 1877, however, the Italian astronomer Giovanni V. Schiaparelli (1835–1910) reported detecting long, narrow dark markings which he called *canali* ("channels").

This was translated as "canals" by astronomers in Great Britain and in the United States, and was taken by some of them to signify the existence of intelligent life on Mars, life that was capable of huge feats of engineering. The strongest proponent of this view was the American astronomer Percival Lowell (1855–1916). He published a book on the subject in 1894 that popularized the notion of Mars as an abode of intelligent life, and H. G. Wells further popularized it with his novel *War of the Worlds,* published in 1898. From then on, Mars was by far the most interesting of the planets to those human beings who were at all interested in astronomy.

Once probes had sent back information on the moon and on Venus, there was serious work in both the Soviet Union and the United States on Mars probes.

The first successful Mars probe was *Mariner 4,* which was launched on November 28, 1964. It passed Mars at a distance of 10,000 kilometers above its surface on July 14, 1965.

As *Mariner 4* passed Mars it took twenty photographs that were beamed back to Earth as microwave signals. For the first time in history it was possible to see the surface of another planet with considerable clarity.

On those photographs, no signs of any canal were to be seen. Rather, a number of craters appeared which were much like those on the moon. What's more, other information relayed to Earth by *Mariner 4* showed the Martian atmosphere to be chiefly carbon dioxide and to be thinner than expected, only a hundredth as dense as Earth's. The chance of intelligent life dropped precipitately.

*Mariner 6* was launched on February 24, 1969, and *Mariner 7* on March 27, 1969. Both were equipped with more sophisticated equipment than *Mariner 4* had been. Their photographs would be clearer and more detailed than those of the earlier probe.

On July 30, 1969, *Mariner 6* skimmed by Mars, and *Mariner 7* followed on August 4. Together, they sent 199 photographs back to Earth.

The new photographs showed that there was no mistake about the craters. The Martian surface was riddled with craters—as thickly, in places, as the moon was. However, the new probes showed that Mars was not entirely like the moon. There were regions in the photographs in which the Martian surface seemed flat and featureless, or jumbled and broken in a way that resembled the surface of neither Earth nor the moon.

The new photographs also showed no signs of canals. Then, too, the probes showed that the temperature of Mars was lower than had been expected, and it didn't seem likely that there was any liquid water on the planet (though it has polar icecaps that might contain ice). With no free water on the surface, no oxygen in the atmosphere, and no sign of canals, the chances of intelligent life on Mars declined to just about zero.

On May 30, 1971, *Mariner 9* was launched. The plan was to have it placed in orbit about Mars, and that was accomplished on November 13, 1971, when it arrived at that planet. It was the first human artifact ever to be placed into orbit about any world other than Earth and the moon (only fourteen years after the first artifact was placed in orbit about Earth).

Shortly after *Mariner 9* moved into orbit about Mars, two Soviet Mars probes arrived, *Mars 2* and *Mars 3*. Each dropped a daughter probe to the surface of Mars, the first on November 27, 1971, and the second on December 2, 1971.

Unfortunately, a planetwide duststorm had started on September 22, 1971. It was still raging when all three probes reached Mars. The Soviet probes that made the first-ever landings on Mars did not survive the storm, but *Mariner 9,* safely in orbit, waited it out.

While waiting, *Mariner 9* took photographs of Mars's two tiny satellites. Up to that time, nothing had been known about them but their orbits and the fact that they were tiny. The photographs showed them to be irregular ovoids, startlingly potato-shaped.

Phobos, the nearer of the two satellites, is the larger, being 28 kilometers across at its widest. Deimos, the other, is 16 kilometers across at its widest.

Phobos is heavily cratered, and the largest crater was named Stickney, after the maiden name of the wife of the astronomer Asaph Hall (1829–1907), who had discovered the satellites. She had urged him to "try one more night" when he was ready to give up.

Deimos has signs of craters, too; the two largest are named Voltaire and Swift after two satirists who spoke of Mars having two satellites long before they were discovered. Its craters, however, seem to be filled, and almost covered, with dust and gravel.

In December 1971, the dust storm finally settled down and *Mariner 9* got to work taking photographs of Mars. In the end, it succeeded in mapping all of Mars so that its surface came to be known as well and as thoroughly as that of the moon.

The first thing that these photographs settled, once and for all, was that there are no canals on Mars. All those dark, narrow lines were optical illusions (as some astronomers had suggested long before the days of probes). The light and dark areas had nothing to do with land and sea, but are the product of light and dark dust distributed by seasonal winds, as the American astronomer Carl Sagan (1935–    ) had suggested a few years earlier.

The craters detected by earlier probes were concentrated for the most part in a single hemisphere of Mars. The other hemisphere has relatively few craters, but has other remarkable features, notably objects that are clearly volcanoes.

The most remarkable volcanoes occur in a region which had been named Tharsis by Schiaparelli. The largest was named Olympus Mons ("Mt. Olympus") in 1973. It is far larger than any volcano on Earth.

Running southeastward from the Tharsis region is a canyon up to 2 kilometers deep, up to 500 kilometers wide, and about 3,000 kilometers long. It is called Valles Marineris ("Mariner Valley"). It is nine times as long as Earth's Grand Canyon, fourteen times as wide, and twice as deep.

What was needed now was an actual soft landing on Mars, such as the Soviets had managed to do in 1971, but it had to be done at a time when no killer sandstorm was in progress on the planet.

With that in mind, *Viking 1* was launched on August 20, 1975, and *Viking 2* on September 9 of that year. Both moved into orbit about Mars. It took a while before a site smooth enough to serve for soft landing was found, but on July 20, 1976, a subprobe of *Viking 1* (a "lander") came down 22.27 degrees north of the Martian equator. Some weeks later, the *Viking 2* lander came down safely 47.97 degrees N of the Martian equator.

As they were landing, the landers analyzed the Martian atmosphere and found that it was 95 percent carbon dioxide, 2.7 percent nitrogen, 1.6 percent argon, and other minor components. The surface temperature was found to be quite cold. Even at the warmest, the temperature of the soil did not rise to the melting point of ice.

Once the landers were on Martian soil, they took photographs that were beamed back to Earth. Both landers, though widely separated, showed the same kind of landscape—a desolate desert area strewn with rocks of all sizes from small pebbles to boulders. The photographs were in color, and they showed the rocks to be a distinct red. The sky appeared to be a bright salmon pink, no doubt from the reddish dust in the air.

Samples of Martian soil were analyzed and they turned out to be richer in iron and poorer in aluminum than Earth's soil is, but the differences were not startling.

Chiefly, however, interest centered on experiments that the lander could perform that were designed to demonstrate whether microscopic life was present in Martian soil. This depended on whether a sample of soil could carry out certain fundamental reactions that Earth life could, under similar circumstances, and, if so, whether the ability of the Martian soil to do so would be lost if the soil were heated under conditions that would destroy Earth life if that were present.

Three separate experiments were tried, and in all three the results were what would be expected if life was present, and yet the results were not so unmistakable as to force that conclusion. This was especially true because a fourth experiment showed that there was no detectable organic material in Martian soil, and, on Earth at least, all forms of life without exception are built up of organic materials. Scientists were not ready to accept nonorganic life, and so they tried to work out ways in which Martian soil might give these lifelike reactions even though no life was present.

The question of microscopic life on Mars is therefore not yet entirely settled, though most scientists would concede that the probabilities seem to be against life.

# 11
# Mars and Beyond

## SPACE SETTLEMENTS

By 1980, then, humanity had thoroughly explored all the large worlds of the inner solar system—Earth and its satellite, the moon; Venus; Mercury; and Mars and its two small satellites, Phobos and Deimos.

There are small bodies that occasionally make temporary appearances within the volume of the inner solar system: asteroids, meteoroids, comets. These may be considered as within humanity's present horizon, for anytime that human beings wish, they can, for instance, set up a probe to study Halley's Comet on its next appearance (in 1986) within the inner solar system.

Again, scientists might set up a probe to study the small asteroid Icarus when it next approaches fairly close to Earth. (This asteroid, not more than a kilometer in diameter, can approach, at times, to within 6,400,000 kilometers of Earth.)

It would be particularly interesting if on one of Icarus's nearer passages it was outfitted with appropriate instruments adequately protected against heat and radiation, for Icarus in its orbit approaches to within 28,400,000 kilometers of the sun. That is a considerably closer approach than Mercury or any other known body (except for a handful of comets) ever makes to the sun. In other words, we might convert Icarus into a "sun probe." (We must remember that the sun, too, is a member of the inner solar system.)

To be sure, all the exploration of the inner solar system, outside the Earth-moon system itself, has been carried out by uncrewed probes, but we might consider that a rather unimportant quibble.

After all, the exploration by such probes has been good enough. The map of Mars is better known today than the interior of Africa was a century ago.

And yet we can't keep from asking the question: Will human beings ever set foot on the worlds beyond the moon?

In this connection, we must remember the matter of distance. Mars is the next logical target after the moon, since Venus and Mercury are too hot and too near the sun to be worth an approach before Mars is. However, a round trip to Mars (or to Venus or Mercury, for that matter) is sure to take over a year, and that would prove a formidable task indeed.

Sending instruments that do not require life-support systems, under conditions in which a miscue that ends in destroying the instruments is disappointing but not tragic, is expensive enough, and failures such as those of *Mars 1* and *Mars 2* are a not inconsiderable economic drain. Add the necessity, however, for supplying all that is needed to support life for a year or more and doing that with an adequate margin of safety and with fail-safe considerations, and the expenses (and emotional investment) will multiply fearfully.

Then, too, the psychological stresses involved in placing a few human beings in lonely confinement with stunted life patterns for many months, together with the knowledge that no rescue is possible if anything goes wrong, may be too much to bear. If we think of that the chance of sending people to Mars would seem to recede.

Arguing optimistically, we might point out that the first of the great sea voyagers—the Phoenicians, Vikings, Polynesians, and Portuguese—all ventured into a vast open space in cramped quarters for many months during which time they might face destruction in any storm without hope of rescue. Yet they managed. If astronauts do as well as the navigators, we face happy success.

We might also argue that astronauts crossing the gulf of space to Mars could keep radio and even television contact with home over much of the trip and would therefore not have the fearful sense of isolation that the early navigators must have had.

Against all that is the fact that the navigators might, and often did, come across land unexpectedly in the course of their journeys. They had the chance to pause for an intermission, so to speak, take on fresh water and other supplies, meet human beings. Astronauts cannot count on any of this.

My guess is that astronauts, taking off from Earth, can and will reach Mars and return, successfully braving and surviving the long, lonely voyage, if some nation, or perhaps an international consortium, is willing to undertake the expense. It would, however, be a *tour de force*, and would not be repeated for a long time. (The first circumnavigation of Earth by the survivors of Magellan's expedition was not followed by a second circumnavigation for some sixty years.)

That is. however. if we restrict ourselves to astronauts taking off *from Earth.*

I have already mentioned the prospect of the development of Earth's near space—that is, the space between itself and the moon. If this is done—if near space is busy with the construction of power stations, observatories. laboratories, and factories—then it is not likely that the work will be done by people who must commute from Earth.

It seems logical to suppose that the builders will remain in space for extended periods of time—especially if most of the construction material must be obtained from the moon. Obtaining the ore will require a mining base on the moon, and it would seem to make sense to have the refining process—the production of metals, oxygen, concrete. soil. grass, and other materials—carried out in space.

In 1974. the American physicist Gerald K. O'Neill (1927-    ) published his suggestion that it was technologically feasible to build large settlements in space capable of housing ten thousand or more people: that these settlements could be designed to have earthlike environments with sunlight yielding an ordinary day-night rhythm; that a centrifugal effect on the inner surface of a rotating settlement could substitute for the gravitional pull on the outer surface of Earth: and that such settlements could make use of the sun as a source for copious supplies of energy.

Since then. the problem of such space settlements has been extensively studied. and it has been pointed out that however expensive the first few settlements might be. the settlers themselves would take over the task of constructing later ones out of matter and energy obtained outside Earth (for the most part) and that the benefits to Earth that would result would quickly far repay with enormous interest any conceivable initial investment.

What's more, present-day technology would suffice to begin the project. Expected advances in technology would make it possible to build settlements more quickly, more cheaply, and more elaborately, to be sure. but present-day technology would suffice.

Standing in the way of such developments at the present moment are economic factors (the worry about the expense), political factors (the desire to spend money on armaments instead), and psychological factors (distrust of other nations, and disbelief in the usefulness of space) but *not* technological factors.

An optimistic view of the future would indicate that before long, the clear necessity of expanding humanity's horizons would cause the space settlements to be built. The construction would also serve as a great project that not only would be clearly of great benefit, but might induce human cooperation in something large enough to fire the heart and mind, and make people forget the petty quarrels that have engaged them for thousands of years in wars over insignificant scraps of earthly territory.

If that should be the case, then the problem of making the long trips to other planets would be completely transformed. Human beings who had lived their lives on space settlements, who had perhaps been born there, would be sure to have an attitude toward spaceflight totally different from that of people who have lived out their lives on Earth.

First, while to Earthpeople spaceflight is something exotic and strange, it would be mundane and everyday to the space settlers. Settlers would routinely work in space, trade by way of space, be constantly aware of space. While space might be an alien environment to Earthpeople, it would be home to the settlers.

Second, while a spaceship offers an environment extremely alien to Earthpeople, that same environment would be homelike to settlers. The settlers would already be living *inside* a world, would be accustomed to an engineered environment and to tight cycling. A spaceship would be smaller than a settlement but it would be of the same species.

Third, while Earthpeople are accustomed to a constant and sizable gravity, settlers would be accustomed to a pseudo-gravity (the centrifugal effect) that would vary with position inside the settlement, and also to the zero gravity that would accompany any work done in a free-fall orbit. The gravitational anomalies that accompany spaceflight would affect the settler far less, psychologically and physiologically, than it would Earthpeople.

Fourth, one of the major difficulties of spaceflight from Earth's surface is the need to overcome the pull of Earth's gravitational field. For space settlements, the pull of Earth would be considerably less strong, and so spaceflight from the settlements would be easier and less costly in energy by far. That, in turn, would mean that it would be practicable to build much larger and more elaborate spaceships, and this too would decrease the tensions of long spaceflights.

Consequently, it may be supposed that when the time comes for human beings to reach out to the planets and for the human horizon to undergo a new major expansion, it will be the settlers and not Earthpeople themselves who will lead the way. It will be the settlers who will be the Phoenicians, Vikings, Polynesians, and Portuguese of the twenty-first century.

The first goal would, as I have said, be sure to be Mars, since it is a world which would be no less exploitable (save for distance) than the moon. In fact, Mars would be more comfortable. Its low temperatures would not be much worse than those of the lunar night, and it would be spared the worst of the solar radiation by day. It has a higher surface gravity than the moon (0.4 $g$, as compared to 0.17 $g$), and it has at least a thin atmosphere as well as some of the volatile elements lacking on the moon.

It is doubtful, perhaps, that the settlers would actually live on Mars, any more than they would live on the moon. Presumably, they would build settlements in orbit about Mars and live there, using Mars itself only as a mining world.

It is less likely that the settlers will reach out in the other direction, toward the sun. Venus would seem to be a completely useless world in terms of human development—impossibly hot at all times, and with an atmosphere unusably dense.

Mercury is less inhospitable. Its nightside would offer no insuperable difficulty to settlers accustomed to the lunar and Martian night, but, of course, any penetration of that nightside would only be temporary. Sunrise would set the limit for any work done on the surface. It might be possible, however, to use the first nightside stay to burrow deeply enough under the surface to find a relatively cool level through which one could survive the surface heat of the 1,400-hour daytime period. Yet even that seems dubious, at best.

There are also discussions, now and then, of "terraforming" planets—that is, of so altering them as to make them more comfortable for human beings. Even if this cannot be carried through to the extent of making planets actually livable, it might serve to make them more useful as a resource base.

The chances are, however, that the initial trend of human exploration and of the expansion of the human horizon would be away from the sun and not toward it. There is much more room away from the sun than toward it, and it is easier to stay warm in a cold environment than to stay cool in a hot one.

Mars, for instance, would only be a way station for what lies farther away. Beyond Mars, there is the asteroid belt with many tiny worlds—perhaps 100,000 worlds with a diameter of more than a kilometer.

These would offer a rich source of ore of all kinds. The total mass of the asteroids is far less than the total mass of the moon, but only the surface layer of the moon is available for exploitation, almost all of it (as almost all of Earth) being too deep for easy reach. Except for the very largest asteroids (Ceres is 1,000 kilometers in diameter, and there are a few dozen asteroids with diameters in excess of 100 kilometers), all the material of an asteroid is so near the surface as to be available for use. Then, too, the asteroids, we are fairly certain, contain the volatile materials that the moon lacks.

The time may come, then, when the asteroids will represent the great mines of humanity. Not only that, but new settlements will be built in the asteroid belt where the material for the purpose exists and where solar radiation will still be strong enough to serve as a useful source of power.

The asteroid belt offers a volume of space perhaps a billion times as great as that which the lunar orbit offers, and one can well imagine that the number of settlements it can hold would be enormous. It is not so difficult to see that the time may come when the vast majority of the human species will be living in the asteroid belt.


## JUPITER

The asteroid belt is the boundary of the inner solar system. Beyond it, away from the sun, are the vast reaches of the outer solar system. In the outer solar system, the worlds are more sparsely spread, but they are larger, too.

There is giant Jupiter, which contains three-fifths of all the mass of the solar system, if we leave the sun itself out of consideration. Beyond it lie Saturn, Uranus, and Neptune, all considerably smaller than Jupiter, but considerably larger than Earth. Jupiter has four satellites that are roughly as large as the moon, or larger: Io, Europa, Ganymede, and Callisto. Saturn and Neptune have one large satellite apiece—Titan and Triton, respectively.

There are also smaller satellites. Jupiter and Saturn have a dozen or so each. Uranus has at least five (but no really large one), and Neptune has one that we know of. In addition, there are small bodies in the outer solar system that are not associated with the four giant planets, and we now know of three: Charon, Pluto, and Chiron.

The distances involved in the region of the giant planets dwarf those in the inner solar system. Jupiter and its satellite system lie at the outermost rim of the asteroid belt, but Saturn lies 650,000,000 kilometers beyond Jupiter's orbit, or 11.5 times the closest approach of Mars to Earth. Uranus lies 2,100,000,000 kilometers beyond Jupiter's orbit, Neptune 3,700,000,000 kilometers beyond, and Pluto

5,100,000,000 kilometers beyond, on the average. The distance from Jupiter to Pluto is six times the total diameter of the inner solar system.

Yet rocket probes have already penetrated beyond the asteroid belt, and have done so successfully.

On May 2, 1972, *Pioneer 10* was launched, leaving Earth with an initial speed of 14.5 kilometers per second, the fastest speed attained by anything human-made up to that time.

*Pioneer 10* passed safely through the asteroid belt, where it detected fewer particles than astronomers had expected in that region, and then reached the vicinity of Jupiter on December 3, 1973. It passed only 135,000 kilometers from Jupiter's surface.

Jupiter's magnetic field turned out to be about forty times as energetic as Earth's and made itself felt at a distance of 7,000,000 kilometers from the planet.

*Pioneer 10* was followed by *Pioneer 11*, which was launched on April 5, 1973, and which passed within 42,000 kilometers of Jupiter on December 2, 1974. *Pioneer 11* was placed on a course that took it over Jupiter's polar regions.

From the information sent back by these probes, it would seem that Jupiter is essentially a vast sphere of hydrogen, gaseous at the very outside, liquid within, and in metallic solid form at the core. There is an admixture of helium, the hydrogen-helium ratio being about 5 to 1, plus smaller quantities of heavier elements. There may be a small, rocky core at the very center.

The planet is hot. At 1,000 kilometers below the visible surface, the temperature is already 3,600° C. (6,400° F.), and at the center of Jupiter, it may be about 54,000° (97,000° F.).

The Pioneers did not get a very close look at the satellites of Jupiter, but this was corrected by another pair of Jupiter probes, *Voyager 1*, launched September 5, 1977, and *Voyager 2*, launched August 20, 1977. Though *Voyager 2* was earlier launched, it traveled more slowly en route and arrived later. *Voyager 1* passed Jupiter on March 5, 1979, and *Voyager 2* on July 9, 1979.

*Voyager 1* detected a ring of debris about Jupiter, essentially like that around Saturn, but much thinner and less extensive, so that Jupiter's ring is quite invisible from Earth. The ring presumably consists of small particles, but there are two or three larger objects associated with it, or right outside it, objects large enough to be thought of as individual satellites that are closer to Jupiter than any of those discovered by Earth-based instruments. They may serve to keep the debris confined within the borders of the rings.

*Voyager 1* photographed each of the four large satellites of Jupiter. In order of increasing distance from the planet they are Io, Europa, Ganymede, and Callisto. Of these, Europa, the smallest, is slightly smaller than our moon, Io is almost precisely the size of the moon, and Ganymede and Callisto are each considerably larger than the moon and are even larger in diameter than the planet Mercury.

Callisto was found to be densely covered with craters, every portion of its surface being part of one crater or another. There are two bull's-eye patterns where particularly large craters splashed, making concentric rings of material.

Ganymede, closer to Jupiter than Callisto is, is also more subject, in consequence, to Jupiter's tidal action. The tidal action can act as a source of heat, and it may be that Ganymede retained its interior heat for longer than Callisto did. As a result, there was still geological activity on Ganymede after the early stage of crater formation. The crust moved and underwent changes, so that although Ganymede has regions of crater formation like Callisto's, there are also regions on the inner satellite's surface where the terrain is clearly younger, and is grooved rather than cratered.

Europa, still closer to Jupiter, is still warmer—warm enough to melt the water that is part of its outer structure. (Ganymede and Callisto are rich in water all the way through, and have the low densities that demonstrate it, but the water is in the form of ice throughout.)

The world-ocean that covers Europa is, however, frozen at the surface, so that the satellite lies under what appears to be a globe-girdling slab of sea ice. The surface of Europa is therefore phenomenally smooth. Any meteor strikes that might have taken place have presumably crashed through the ice and splashed the water beneath. Refreezing would wipe out any sign of the encounter. (There are traces of three craters, however, where ice may have piled up about the strike, without there having been sufficient time, as yet, for a smoothing-out process.)

One more odd thing about Europa is that the surface is covered with straight black markings that may be cracks in the ice cover. These look amazingly like the markings that astronomers once drew on Mars and interpreted as canals.

Io, the innermost of the large satellites, is naturally the one most affected by Jupiter's tidal actions. Three days before *Voyager 1* passed Jupiter, a paper in *Science* predicted that active volcanoes might be found on the satellite as a result. The prediction was a correct one. No less than eight volcanoes in active eruption were detected on Io by *Voyager 1,* and when *Voyager 2* passed, four months later, six of them were still erupting. Io is the only world, other than the earth, on which active volcanoes have been reported.

Io has a density similar to our moon's and is essentially a ball of rock. Water and other volatile materials have been driven off by its internal heat, and like the other large Jovian satellites it has no atmosphere beyond trace wisps of gas. In Io's case, the trace is mostly sulfur dioxide, which is born of the volcanic eruptions that spew dust and ash onto the airless surface. Much of the eruption is sulfur vapor, which freezes, and so Io is covered by a red-and-yellow surface of sulfurous lava that fills in any crater that may have formed. Some craters can be detected, having been formed recently and having not yet been entirely filled in.

## THE OUTERMOST PLANETS

Various Jupiter probes have had their orbits designed in such a way as to make them move on to close passage with Saturn and not long afterward *Voyager 2* followed. They have thus become Saturn probes. *Pioneer 11* passed within 20,200 kilometers of the visible surface of Saturn on September 1, 1979. It sent back detailed photographs of the rings, which showed them to be more complex than it was possible to make out by Earth-based telescopes.

Then, on November 12, 1980, *Voyager 1* passed Saturn. Now the rings were seen in enormous complexity. They are made up of something like a thousand sub-rings. Even the "Cassini division," which, from Earth, appeared to be a completely empty band dividing the two main rings, proved to contain several thin sub-rings of particles within it. The rings extend farther in toward Saturn and outward from it than could be seen from Earth. Some of the sub-rings seem not quite symmetrical and one even seems braided. As in the case of Jupiter, small satellites (at least three of them) exist along the edges of the rings, and serve, perhaps, to confine the particles within the rings.

A number of the satellites were photographed. Most had craters, as was to be expected. Mimas had a very large one in relation to its own size. The collision that formed the crater may very nearly have shattered the satellite.

The satellite Dione was found to have a second, smaller satellite sharing its orbit and separated from it by 60 degrees. Dione and its companion mark the vertices of an equilateral triangle with Saturn at the third vertex. This is called a "Trojan situation" and is the first to be found involving a planet and two satellites. Jupiter, the sun, and two groups of asteroids form the vertices of a pair of adjacent equilateral triangles. It is because the asteroids had been given the names of heroes of the Trojan War that the configuration gained its name. The Italian-French astronomer Joseph Louis Lagrange (1736–1813) first showed, about 1790, over a century before any examples had been discovered, that such a configuration was gravitationally stable.

The satellite Enceladus was found to be very smooth, rather like Jupiter's Europa, but *Voyager 1* did not get a really good look.

Iapetus is another curiosity. Even from Earth, it was noted to be much brighter when on one side of Saturn than when on the other. Since astronomers assumed it always faced one side to Saturn, that meant that one hemisphere was visible on one side of Saturn and the other hemisphere was visible on the other side of Saturn. The conclusion was that Iapetus, for some reason, reflected far more light from one hemisphere than from the other. *Voyager 1* showed this to be true, and Iapetus showed up clearly as a "two-tone" satellite. One half of it is probably ice-covered and the other is bare rock—but why?

The prize of the system, though, is Titan, Saturn's one huge satellite, one that rivals Jupiter's Ganymede in size. Even from Earth, Titan had been found to have an atmosphere—the only satellite in the solar system known to have one that contained more than ultrathin traces of gas. What's more, a particular gas, methane, had been located in its atmosphere.

Methane is made up of molecules containing a carbon atom and four hydrogen atoms. It is the simplest of the organic molecules. There were speculations that on Titan's surface there might be more complicated molecules of the same sort so that it might be a world of gasoline oceans and asphalt continents. Even more dramatic was the thought that life is based on organic molecules and that Titan might possess life forms of some sort. Astronomers hoped that *Voyager 1* might answer questions in this connection.

Titan proved indeed to have an atmosphere, and a far denser one than had been expected—at least 1 1/2 times as dense as Earth's. However, the great density was due to nitrogen, a gas not easily detected from Earth. Methane was present as a major impurity. The surface was cold enough for the satellite to have lakes of liquid nitrogen with, perhaps, methane in solution. Life forms might still exist if it is possible that they could survive liquid-nitrogen temperatures. Unfortunately, there was no way of seeing the surface, for the Titanian atmosphere possessed a haze as impenetrable as Venus's clouds. Microwaves will be required to penetrate it.

So far (as I write), probes have not sent back information from beyond Saturn, though at least one of the Voyagers may eventually reach the vicinity of Uranus and still be operational. From Earth-based instruments, however, astronomers have managed to learn a few new things about the outer solar system in the last few years.

On March 10, 1977, the American astronomer James L. Elliot, observing the occultation of a star by Uranus, found that the star flickered several times in a certain pattern as it approached Uranus, and then flickered in precisely the reverse pattern after it left Uranus. The best interpretation of that is that a number of thin rings of particles surround Uranus.

There are thus three ringed planets—Jupiter, Saturn, and Uranus—and it will not be in the least surprising, now, if it turns out that distant Neptune has rings too. Circling debris may be as natural an accompaniment of large planets as a family of satellites is. What *is* unusual, therefore, is not that Saturn has rings, but that it has so enormous and brilliant a set of rings.

On November 1, 1977, the American astronomer Charles Kowall discovered an asteroid whose orbit lay between that of Saturn and Uranus. He named it Chiron. This may herald the existence of a whole family of asteroids in the outer solar system. What we call the "asteroid belt" may only be the closest and, therefore, the most easily detected of several such.

Even Pluto had a revelation to make. It, the most distant planet, was discovered in 1930 by the American astronomer Clyde William Tombaugh (1906–    ). Originally, it was thought to be a sizable body, as massive as Earth (though by no means a giant like the outer planets). However, the more it was studied, the more reason there seemed to be to consider it a smaller body. With each estimate, its mass seemed to shrink.

Pluto has a more elliptical orbit than any of the other planets and, at perihelion, it is actually closer to the sun, for a period of twenty years, than Neptune is.* Right now, Neptune is near perihelion, and is more easily viewed from Earth than it will be throughout the twenty-first and twenty-second centuries.

On June 22, 1978, then, it was possible for the American astrono-

---

* When the solar system is shown on a flat diagram, the orbits of Neptune and Pluto seem to cross, but actually the orbits are tilted to each other and Pluto is far distant from Neptune in the third dimension at the "crossing" points.

mer James W. Christy to just make out that Pluto had a satellite almost as large as itself. He named it Charon.

From the distance between Pluto and its satellite, and from the period of revolution of the satellite, it was finally possible to work out the mass of Pluto. This turned out to be only 1/8 the mass of our moon, so that Pluto is far smaller than anyone had thought. Pluto is about 3,000 kilometers in diameter and Charon is about 1,200 kilometers in diameter. These worlds are no larger than satellites of moderate size.

The discoveries of the last decade, then, whether through probes or otherwise, have made it clear that the outer solar system has a lot more to it than we had expected, and that continued exploration would yield much of interest.

Are we condemned, however, to carry through this exploration by means of ever more sophisticated probes, or is it conceivable that someday human beings will themselves penetrate the outer solar system?

Again, if we consider such explorations as starting from Earth, it doesn't seem likely. Space probes take eighteen months to pass from Earth to Jupiter, and an additional twenty months to reach Saturn. Clearly a round-trip exploration of even the nearer worlds of the outer solar system is going to take the better part of a decade.

However, once the asteroid belt becomes part of the human domain and includes many populated settlements, the problem becomes less acute. Some of the settlements may, at times, be reasonably close to Jupiter and could make the trip to Saturn. Asteroids beyond Jupiter (after the fashion of Chiron) may exist, and if bases are established there, or on the outermost satellites of Saturn, voyages beyond Saturn may be ventured upon.

To be sure, it is not likely that human beings will venture to penetrate the atmosphere of the four large outer planets themselves in the foreseeable future. What's more, the most interesting worlds of the outer solar system other than the giant planets themselves—the four large satellites of Jupiter—may be unapproachable without as yet unforeseen technological advances. Those satellites circle within the magnetic field of Jupiter, and the radiation intensity therein would be deadly to the crews of spaceships, as now constructed.

Nevertheless, there are at least eight small satellites of Jupiter circling it outside its magnetic field, and observatories could be placed upon them for the study of Jupiter and the inner satellites until human beings learn a technique for approaching them more closely.

## COMETS

Even if the entire planetary system finally falls within the human horizon, we are not yet necessarily through.

The most distant planetary object we know is Pluto. At the moment, as I said earlier, Pluto is near its perihelion and is closer to the sun than Neptune is. Pluto, however, will then recede over the course of the next century and a quarter, reaching the farthest point from the sun, its aphelion, about 2114. It will then be about 7,350,000,000 kilometers from the sun, or 1.6 times as far as Neptune ever gets.

But is there any object in the solar system that we know of with an orbit that carries it beyond Pluto?

Yes! There are objects we occasionally see which have orbits so elongated that though they may be quite near the sun at perihelion, they move far beyond Pluto's distance from the sun at aphelion. These are the comets.

To be sure, some of the comets we see have been captured by the gravitational attraction of the planets (particularly that of Jupiter) so that even at aphelion they are closer to the sun than Pluto is.

Other comets, however, which at one end of their orbit lie within the planetary system will at the other end be 1,000,000,000,000 kilometers or more from the sun. We can imagine that instruments placed on such a comet when it is within reach would be carried far out into trans-Plutonian space and record conditions in distant space for the million years or so that will elapse before the comet reenters the planetary system—assuming, of course, that the instruments will endure, record, and transmit for such a period of time and, for that matter, that the human species itself will endure.

Of course, there may be many comets that never enter the planetary system but, through all the 4,600,000,000-year history of the solar system, have been circling the sun, unseen and as yet unseeable at their vast distances.

Astronomers believe that there may be as many as 100,000,000,-000 comets which revolve in these very distant orbits, orbits that at no point come within a trillion kilometers of the sun.

Those venturesome comets that visit us here in the inner solar system arrive because their particular orbits were perturbed by collisions with each other, or by the gravitational influences of the sun's neighbor stars. This rarely happens, and through all the solar system's long history, the distant comet cloud has been depleted of no more than one-sixth of its original content, at most.

The comets reaching the inner solar system become spectacular sights as the heat of the sun warms their frozen structure to release a haze of vapor and dust, which is then swept outward into a long tail by the solar winds—the stream of high-speed electrically charged subatomic particles forever being ejected from the sun in all directions. The effect of heat and the solar winds erode the comet, drive off its volatile components, and leave a residue which may not be representative of its original composition. Yet the original composition may be the most nearly unchanged remnant of the vast nebula of dust and gas out of which the solar system was formed. It might, therefore, seem valuable to study distant comets and gain an insight into the beginnings of the solar system.

This does not mean, however, that we need venture trillions of kilometers beyond Pluto to do so. There may well be comets that enter the planetary system but whose orbits have perihelia that are located well beyond the orbit of Saturn. They would never be seen from Earth, but might be located and studied if we were exploring the outer solar system. Such comets, located far from the sun at best, are not heated appreciably and would continue to have essentially the composition they have had from the time they were formed.

It might well be that people might decide there was no use in venturing trillions of kilometers beyond. Pluto merely to find more examples of comets exactly like those they could find in the region of the outer planets.

If, then, there were nothing permanently beyond Pluto but a vast number of comets, the rule of "See one and you've seen them all!" might keep humanity from venturing farther, and Pluto might forever mark the outermost limit of humanity's horizon.

# 12
# The Stars

## THE DISTANCE OF THE STARS

As it happens, a human horizon confined to the solar system would be limiting indeed, for there is a great deal beyond Pluto, and even beyond the comets. Outside the solar system stretches the vast universe of the stars.

Already, humanity is reaching for those stars. *Pioneer 10,* the first of the Jupiter probes, is heading out of the planetary system altogether. As this book is being written it is in the neighborhood of the orbit of Uranus. In another three years, it will be passing Neptune's orbit and will be heading outward indefinitely.

With *Pioneer 10,* as it goes out into the unknown, there will be a message from Earth etched into a 15-by-22-centimeter gold-covered aluminum slab. The message was designed by Carl Sagan and Frank D. Drake (1930–    ), and was drawn by Linda Sagan.

The most noticeable thing about the message is the outlined figure of a man and a woman, unclothed, with the differences in sex discreetly indicated, thus giving certain minimal information about the kind of creatures that built *Pioneer 10* and sent it on its way. The man is holding up his hand in what is hoped will be taken as a gesture of friendship and peace. If not, it at least shows the existence of four fingers and a thumb.

Behind the man and the woman is an outline of *Pioneer 10* to scale. If intelligent beings find *Pioneer 10* someday and measure its dimensions, they will know the size of human beings as well.

At the bottom of the slab are circles representing the sun and its nine planets, giving some indication of their relative sizes, and of the rings about Saturn, together with a line marking the path of *Pioneer 10* among the planets. That should be enough to identify the solar system as the place of origin of the probe.

Other symbols are included which express the location of the sun in the galaxy, and which give an idea of our scientific advancement.

*Pioneer 11,* which will also leave the solar system, carries a duplicate of the plaque.

As for the Voyager probes, they appeal not only to sight, but to sound as well. In addition to carrying photographs of representative scenes on Earth, they carry a recording of representative sounds.

It is very unlikely that any intelligent beings will ever find these small, wandering messages in the infinity of space—and if such a find is ever made, it will almost certainly be millions of years hence. Nevertheless, it seems typical of humanity that it should wish to advertise its existence and accomplishments, just in case, and it is perhaps a rather admirable and justifiable pride.

But why should people wish to send the messages out beyond the solar system? What is out there?

The very notion that anything at all is out there is a relatively modern one. Throughout ancient and medieval times, the stars were viewed rather as though they were luminous sparks attached to the dark, solid hemisphere of the sky that wheeled around Earth, not very many kilometers above the peaks of the higher mountains. The sky was Earth's solid boundary, a mere adjunct, and not a universe.

Even after the scale of the solar system was determined by Cassini in 1672, and it was understood that Saturn (then the farthest known planet) was some 1,400,000,000 kilometers from the sun, and that the stars must be farther still, that did not improve matters. It was still possible to think of the stars as painted on a solid sky that was not much beyond Saturn. The sky became the boundary of the solar system, but it was still not a universe.

One person who took a contrary view lived more than two centuries before Cassini. He was a German scholar named Nicholas of Cusa (1401-64) who, in a book published in 1440, maintained that space was infinite and that the stars were suns strewn through that infinite space in infinite numbers. These other suns, he maintained, were all surrounded by planetary systems, and so there were an infinite number of inhabited worlds.

Nicholas of Cusa had no evidence for these views, but maintained them out of some inner intuition, feeling they made sense. These views must, at the time, have seemed eccentric and many scholars must have smiled indulgently over them. Nicholas's career does not seem to have been damaged, however, for in 1448 he was made a cardinal, and he lived and died in peace.

His views were taken up a century and a half later by the Italian scholar Giordano Bruno (1548–1600), but by this time, the Protestant Reformation had taken place. What's more, Copernicus had published his book placing the sun at the center of the planetary system, and the Catholic Church felt itself to be fighting for its life against both scientific and religious rebellions. There was less tolerance for what seemed eccentric opinions, and Bruno was burned at the stake.

Yet by Bruno's time there was at least negative evidence in favor of great distances to the stars.

When Copernicus announced his theory, there were those who pointed out that if Earth really moved through space and revolved about the sun, then astronomers would be viewing the stars from radically different regions of space, depending on whether Earth was at one end of its orbit or the other. This shift in position would produce a "parallax" in the stars (if Nicholas of Cusa's suggestion was correct). Those stars that were nearer would seem to shift their apparent position in the sky, relative to those which were farther, as Earth moved in its orbit.

This was a good point and was sufficient, all by itself, to show that either Copernicus or Nicholas of Cusa was wrong (or both, for that matter) unless there was some way to explain the absence of parallax.

There *was* a way. Copernicus saw that way and used it. He pointed out that parallax grew smaller with distance and that it must be that the stars were so far away from Earth that the entire shift in the position of even the nearest, as our planet moved through its orbit, was insignificant. In short, there *was* a parallax among the stars, but thanks to their great distances, it was far too small to measure.

That seemed almost as ridiculous a notion, to the traditionalist, as Earth's motion about the sun was, but as the Copernican theory grew more and more useful, the deduction therefrom that the stars were very far away seemed more and more acceptable.

The English astronomer Edmund Halley (1656–1742) harked back to the notions of Nicholas of Cusa, about 1718, in order to make the first scientific estimate of the distance of the stars. Suppose the stars were indeed suns and suppose their insignificant light in comparison to our own sun was entirely due to their vast distances. Consider Sirius, the brightest star in the sky. Suppose it was actually a body exactly like the sun in size and brightness. How far away must it be in order to shine with no more than its apparent brightness in the sky, the brightness of a tiny star?

Halley estimated that Sirius's distance from Earth, in that case, was 125,000 times the sun's distance from Earth. Using the modern value of the distance of the sun from Earth (rather than the somewhat lower value available to Halley), this would mean that Sirius was 19,000,000,000,000 (or $1.9 \times 10^{13}$) kilometers from us.

That is a huge distance—19 trillion kilometers—and not easily grasped. One dramatic way of expressing it is to make use of the speed of light.

The first useful determination of the speed of light was made in 1676 by the Danish astronomer Olaus Rømer (1644–1710). He used, for the purpose, the manner in which the eclipses of the satellites of Jupiter were delayed as Earth receded from Jupiter (as each moved in its orbit) and advanced as Earth approached. The value he obtained was only three-quarters of the value that is now accepted, but it was excellent for a first attempt. The value of the speed of light, as accepted today, is 299,792.5 kilometers per second.

This is an enormous speed by earthly standards. In one year, light, traveling through the vacuum of outer space at this speed, will cover 9,460,563,614,000 (or just about $9.46 \times 10^{12}$) kilometers, and that distance is called a "light-year."

A notion of the size of a light-year can be obtained if we understand that the diameter of the orbit of Pluto, the most distant planet, is just a little over a thousandth of a light-year.

Yet Halley's estimate placed Sirius at a distance of just about two light-years from us.

Of course, Halley's estimate depended on Sirius's being as bright as the sun, which might not be so, so his estimate had to be taken with great caution. To obtain a more reliable estimate of distance, one would have to measure the actual parallax of a star, and this was too small to be measured by the telescopes that were available in Halley's day.

The feat was not accomplished till 1838, when the German astronomer Friedrich Wilhelm Bessel (1784–1846) announced the determination of the parallax of the dim star 61 Cygni. Soon afterward, the Scottish astronomer Thomas Henderson (1798–1844) announced the determination of the parallax of the bright star Alpha Centauri, and the German-Russian astronomer Friedrich von Struve (1793–1864) announced that of the bright star Vega.

It turned out that Alpha Centauri (actually a system consisting of three stars) was closer to us than any other star in the heavens, and even so, its distance was 4.3 light-years. The star 61 Cygni, is 11.2 light-years away, and Vega is 27 light-years away. Sirius, as it turns out, is about 23 times as luminous as the sun and would therefore have to be much farther away than Halley thought to be as dim as it is. Sirius is 8.16 light-years away.

Suppose, then, we have perfected our spaceflight capacities to the point where we can travel across the solar system without too much difficulty, having learned to make trips of up to 9,000,000,000 kilometers in a few years in large, comfortable spaceliners. Such a distance is, however, only 1/4,500 the distance to Alpha Centauri, the nearest star. If we could cross Neptune's orbit from end to end in two years, that same average velocity would not allow us to reach even the nearest star in not less than 9,000 years.

As it happens, there are many stars visible in our sky that are hundreds of light-years distant, and even these are our neighbors. The sun and all the stars in the sky visible to us without a telescope are part of a huge pinwheel-shaped structure called the "galaxy" and that is 100,000 light-years across. There are other galaxies, too, each with anywhere from a billion to a trillion stars, that are millions of light-years away and even billions of light-years away.

## THE LIGHT BARRIER

It seems useless to speak of those enormously distant stars when even the nearest stars would seem to be out of reach, and yet there remains an enormous attraction about those far-off objects.

In recent years, astronomers have learned a great deal about the universe they didn't know before, thanks to radio telescopes and to satellites that can pick up radiation, such as X rays, which is absorbed by our atmosphere and is therefore impossible to observe from the surface of Earth.

We know about objects such as quasars, pulsars, black holes, and exploding galaxies, which no one had ever heard of a quarter of a century ago, and many people would dearly love to know more about them. It would be in the highest degree dangerous to approach such objects too closely, of course, but we could approach a *little* more closely to them than our present location without being harmed.

Then, too, do other stars have planetary systems? Are there planets sufficiently like Earth to permit settlement—or to bear life of their own? We have not found life in our own solar system, except on

Earth, and may not do so in the future, but our solar system is but one of uncounted trillions, perhaps, and why should there not be life elsewhere? Intelligence, too, and even civilizations!

But though curiosity may burn within us, those vast differences remain to dishearten us.

One's first hopeful thought is that distances tend to shrink as technology advances. It took Magellan's one surviving ship almost exactly three years to circumnavigate the world. It took the astronauts three days to go from Earth to the moon, a distance about 9.5 times the circumference of Earth. The average speed of the first astronauts was therefore about 3,500 times as great as the average speed of the first circumnavigators. Might it not be that with advances in technology we can multiply the speed of our interstellar spacecraft 3,500 times beyond that of our Apollo program?

Reaching the moon in three days means an average rate of progress toward the moon of about 1.5 kilometers per second. Increase that by 3,500 times and we can imagine ourselves traveling at an average rate of 5,250 kilometers per second. At that rate, it will still take about 250 years to reach Alpha Centauri, the nearest star.

Well, then, increase the speed by another 3,500-fold factor, and we will be moving a little over 18,000,000 kilometers per second. It will then take us only four weeks to reach Alpha Centauri.

Unfortunately, that can't be done as easily as one can write it down. In 1905, the German-Swiss physicist Albert Einstein (1879–1955) advanced his Special Theory of Relativity, according to which it is impossible for anything to travel faster than the speed of light in a vacuum. The Special Theory has been checked innumerable times in the eight decades since it was advanced, in innumerable ways, by innumerable people, and it has held fast and firm. No physicist expects to see the speed of light surpassed.

If this is so, that means the fastest any spaceship will ever travel will be 299,792.5 kilometers per second, and at that rate, it will take 4.3 years to reach the nearest star, 30,000 years to reach the center of our galaxy, 300,000 years to circumnavigate it, 2,300,000 years to reach the Andromeda galaxy, 1,000,000,000 years to reach the nearest quasar, 10,000,000,000 years to reach the farthest quasar we have detected, and perhaps 40,000,000,000 years to circumnavigate the universe.

In short, the speed of light, which is unimaginably fast by earthly standards, is an impossibly slow crawl in terms of the universe as a whole, and we are condemned to crawl just that slowly, or more slowly still. Anything but a few neighbor stars is really out of reach even if we are prepared to spend a lifetime traveling.

But can we be sure? Some people who don't really understand physics feel we will somehow find a way to "break the light barrier" if we only use enough rocket push.

That, unfortunately, is not so. Acceleration increases the energy of motion, or "kinetic energy," but there are two factors involved in that energy: velocity and mass. At low speeds, almost all the increase in energy involves velocity, so that an object goes faster and faster, while the mass gains only an undetectable amount. As the speed increases, however, less and less of the further increase of kinetic energy goes into velocity and more and more into mass. As the speed of light is approached, almost all the increase in kinetic energy goes into mass, so that the mass grows larger and larger while the velocity scarcely changes. Increasing the force of the rocket engine, and continuing to do so with the utmost persistence, will merely balloon the mass farther and farther upward toward the infinite without ever affecting the velocity sufficiently to push it past the speed-of-light barrier.

It may strike you that that's not the way the universe should be, but that's the way it *is!*

To be sure, Special Relativity applies to objects we know. Objects with mass, such as ourselves and our spaceships, can travel at any speed (in theory) from zero to the speed of light. Objects without mass, such as light waves, can travel in a vacuum *only* at the speed of light, neither faster nor slower.

But what about objects with an amount of mass that is expressed by what mathematicians call "imaginary numbers"? If such numbers are fitted into Einstein's equations, it would seem that any object possessing such mass must always move *faster* than the speed of light. It can move at any speed from that of light to infinite speed, but never *less* than the speed of light.

The existence of such faster-than-light particles was first suggested in 1962 by the physicists O.M.P. Bilaniuk, V. K. Deshpande, and E.C.G. Sudarshan. Some years later, the American physicist Gerald Feinberg (1933–   ) coined the word "tachyon" (from the Greek word for "fast") for the particles.

Suppose, then, that ordinary particles were converted each into some corresponding tachyon. The tachyons would instantly move off at incredible speed, and if these tachyons were then converted back into ordinary particles at a certain time, enormous distances could be covered in a matter of days, or even seconds.

Unfortunately, no one has yet detected tachyons, and there are many physicists who argue that they cannot exist even in theory. If they could exist, the use of a "tachyonic drive"—the conversion of ordinary particles into tachyons, the control of tachyonic flight, and the reconversion of tachyons into ordinary particles—would all represent difficulties of enormous magnitude. We cannot feel confident that these difficulties can be overcome in the foreseeable future.

Another way of looking at the light barrier is to suppose that it applies only to conditions we have been able to investigate and test. Under conditions far beyond anything we can test, the limit may break down.

There are, for instance, black holes, in which matter compresses to such extremes that density and gravitational intensities approach the infinite. Under such conditions, would the Special Theory hold?

Some astronomers have indeed suggested that objects passing through a black hole may emerge in a far-distant part of the universe in a very short time. In that case, we might look at black holes as each the terminus of a particular "cosmic subway line." If we map each black hole and determine where its other end emerges, we can manage to go from one place in the universe to any other by selecting the proper black holes, and we would be restricted to the light barrier only in traveling from the exit of one black hole to the entrance of another.

But then, not all physicists agree with this view of black holes. And even if black holes are indeed cosmic subways, the matter of finding them, mapping them, discovering them to be sufficiently close to each other for practical purposes, and (most of all) working out a way of entering one without being torn apart by tidal forces would offer difficulties every bit as enormous as those involved in the tachyonic drive.

Again, we cannot confidently predict the use of black holes for interstellar travel in the foreseeable future.

It may be, then, that we will simply have to reconcile ourselves to the existence and the invulnerability of the light barrier.

## BELOW THE LIGHT BARRIER

The light barrier is, in one way, not as bad as it sounds. The Special Theory makes it clear that the experience of time passage slows as speed increases. At first, the slowing effect is imperceptibly small, but as the speed of light is approached, the effect becomes more and more pronounced, until at the speed of light, the experienced rate of time passage is zero.

In other words, if space travelers could go at the speed of light, they would experience no sense of time passage, however far they traveled. They could go from Earth to the farthest quasar, 10,000,-000,000 kilometers away, and it would seem to them that it had all taken place in an instant.

In that case, why bother passing through black holes or converting ships and crew into tachyons? Why not simply convert ship and crew into light waves—a pattern of light waves that would retain within its complexity every detail of the ship and crew? The flight would then be beamed in some desired direction, and the light waves would be reconverted into the ship and crew again. Whatever the destination, it would have been reached in zero time.

Such a procedure, however, offers difficulties as complex and as apparently insuperable as those involving tachyons and black holes. It is certainly not within our grasp in the foreseeable future.

Is there any way of achieving the speed of light (or nearly that speed) without actual conversion of particles into light waves?

What about ordinary acceleration? We cannot accelerate a ship past the speed of light, but we can accelerate it (in theory) to any speed up to that of light.

Acceleration can't be too rapid, of course, since that would smash any crew members against the rear of the vessel. Suppose, though, we accelerate at 1 $g$ (the normal acceleration associated with Earth's surface gravitational field). We could experience that indefinitely and still feel completely comfortable. In fact, we would feel as though we were standing on Earth and experiencing its surface gravitational pull.

At 1 $g$, we go, every second, 0.0098 kilometers per second faster than the second before. If this were to continue indefinitely* then it would take just about a year to progress from zero speed to the speed of light, by which time the ship would have traveled half a light-year.

At that point, the ship could stop accelerating and merely coast. The ship could coast through vacuum indefinitely without slowing or swerving, if there were no nearby stars to exert a substantial gravitational pull. (If there were, this could be taken into account.)

During this coast at very nearly the speed of light, there would be very little sensation of time passage—a few minutes, a few days, a few weeks, depending on just how close to the speed of light the ship was moving and for how great a distance it coasted. Then, when one was half a light-year from the destination, one would decelerate at 1 $g$ for a year.

---

* Actually, it wouldn't, for as the ship went faster and faster, more and more of the accelerating force would go into mass and less and less into velocity. We will simplify matters, however, for the sake of the argument.

Any interstellar flight would then seem to take not much more than two years. Allowing one year for exploration, a round trip to Alpha Centauri would take about five years, and a round trip to some star in the Andromeda Galaxy would also take five years.

This would seem a cheerful situation, except that there are some drawbacks.

First, when a spaceship coasts at, or nearly at, the speed of light, it is only to the crew on board that the experience of time passage seems to slow. To everyone on Earth (and anywhere else in the universe where motion is proceeding at ordinary speeds), time is experienced as passing in the normal manner.

Thus, a round trip to Alpha Centauri might take five years to the astronauts but on returning to Earth they would find that eleven years had passed. This might not bother them too much, but if they had traveled to the Andromeda galaxy and back, they would find that on Earth, 4,600,000 years would have passed, and there might be no human beings to greet them. If they had traveled to the farthest quasar and back in five years, 20,000,000,000 years would have passed on Earth, and the Sun would be a fading white dwarf. Earth would probably have been physically destroyed by then during the sun's red-giant stage.

Even if astronauts understand this and are willing to explore the universe without thought of ever returning to Earth, we must ask ourselves further if it is really practical to expect to accelerate to nearly light velocity.

A year's worth of acceleration at 1 g consumes a vast amount of fuel, more than we can expect a ship to be able to carry, especially since there will be, eventually, a year's worth of deceleration, which will be just as expensive in fuel.

Even using the most efficient form of energy possible, it would take so much energy to accelerate for a year and decelerate for a year that the practicability of carrying the fuel is in serious doubt.

But then, perhaps ships won't carry any fuel at all except for small quantities for emergency purposes. Instead, a ship could sweep up the thin wisps of matter in interstellar space and use that as fuel. It would require a vast "vacuum cleaner" to do this, for it would be necessary to sweep up all the matter in a volume of many thousands of cubic kilometers per second, perhaps.

Then, too, even if the problem of fuel is solved in one way or another (which doesn't sound likely), other problems remain.

For instance, as speed grows closer and closer to that of light, more of the force used to accelerate goes into raising the mass of the spaceship and less into increasing its velocity. A time will come when, whatever the propulsive mechanism, the ship will not be able to afford the waste of using more and more energy for conversion into less and less velocity increase.

Let us suppose, for instance, that the breakoff point comes at 90 percent of the speed of light. That seems pretty good—but it isn't really. At 90 percent of the speed of light, the crew of the ship experiences a time-passage rate that is 31 percent that of normal. That means a trip to the center of our galaxy, for instance, will take up to 10,000 years and a trip to the Andromeda galaxy will take 800,000 years.

If the ship is not able to get that last 10 percent of light speed and reduce the experienced rate of time passage to a really small fraction of normal, then exploration within a normal human lifetime will be confined to the stars of our own neighborhood, those within a couple of hundred light-years.

And even this might not be the worst of it. Interstellar space is not truly empty, and the faster a ship goes, the less able it will be to avoid a collision. It is extremely unlikely that a ship will strike a star, because these are distributed very sparsely indeed. A ship could move very quickly at random and probably not strike any really large body in millions of years.

The number of small bodies in space—flying mountains, boulders, pebbles—must be far greater than stars, planets and satellites. Collisions with any of these smaller bodies, even pebble-sized ones, would, at speeds near that of light, be catastrophic.

Even if there were no debris at all in interstellar space, or if the ship managed to avoid all the debris that does exist, that would not be enough. We know that interstellar space contains dust particles, as well as individual molecules and atoms. We can actually see dust clouds from our vantage point on Earth and identify some of the atoms and molecules within them. Even space that is otherwise clear contains a thin scattering of hydrogen atoms. The clearest space is probably that between galaxies, and it is estimated that out there, there is one hydrogen atom every cubic meter. There is a thicker scattering of hydrogen atoms in interstellar space within a galaxy.

Hydrogen atoms matter. If a ship is moving at a large fraction of the speed of light and strikes a hydrogen atom, that is equivalent to a hydrogen atom moving at a large fraction of the speed of light and striking a ship. Such a speeding hydrogen atom is a cosmic ray.

If there was only one hydrogen atom in every cubic meter, a spaceship going at nearly the speed of light would be exposed to something approaching a billion billion cosmic-ray particles every second. The ship would quickly become radioactive and everyone on board would be fried.

To avoid this radiation danger, it might be necessary to go at no more than one-tenth the speed of light, or perhaps even less in regions that are unusually dusty. At such a speed, the subjective experience of time passage is almost normal and it would take forty years to get to even the nearest star.

## COASTING

It would seem, then, that barring some technological breakthrough that is completely unforeseen now, human beings may never be able to travel at any speed more than a relatively small fraction of light. This, in turn, means that human beings might never be able to make a round trip to even the nearest star in a normal life span.

Is there any way, though, in which a normal life span might be indefinitely extended? What if human beings could be quick-frozen into a state of dormancy? They would then be, it might be hoped, not dead, but merely in a state where metabolism is suspended. If they could be kept so, at liquid-nitrogen temperatures, indefinitely, then when some destination was approached, they might somehow be rapidly warmed and restored to life, and they would then continue without any sensation of time passage having taken place while they had lain dormant.

Unfortunately, having a large, warm object such as a human being so quickly frozen in every part as to prevent degenerative changes that would be fatal would be a tall order indeed. It would be just as difficult to arrange to have every part of a large, very cold body so quickly warmed throughout as to prevent degenerative changes that would be fatal.

Such freezing-unfreezing, then, is not the sort of thing we can suppose will come to pass in the foreseeable future, though it does not seem quite as far out as tachyons, black holes, or the conversion of human beings into light waves.

But suppose the life span of human beings was greatly expanded so that they could undergo long journeys without the need for being frozen. Indeed, if human beings were to attain near-immortality, it might be necessary for them to undertake long journeys, since the solar system would soon be bursting at the seams with its rapidly expanding population.

Greatly expanded life spans are also not very high in probability, though this, perhaps, is less improbable than the successful use of freezing.

It seems fair to conclude, however, that if trips are to be made into interstellar space, they will have to be made at relatively low speeds, without freezing, and with crew members of normal life span. This means, in turn, that the interstellar explorers will have to plan on leading their whole lives on board ship, on having children who will, in turn, spend their whole lives on board ship, and so on for many generations, perhaps.

This may sound, at first thought, like a horrible state of affairs that people could not possibly endure—but we are thinking of Earthpeople of today, used to the special conditions of Earth.

What if we consider the numerous space settlements that may someday fill the asteroid belt and the outer regions of the solar system? Each space settlement will itself be a "starship." Propelled and powered, perhaps, by controlled hydrogen fusion, such settlements could, if they wished, put an end to our forced accompaniment of the sun on its quarter-billion-year orbit about the galactic center. Each could break free and go its own way, moving through space at will.

There would be no great sense of isolation or imprisonment, for the crew would be the people of the settlement, who had been born there and had lived there all their lives, whose parents and grandparents had perhaps lived there, whose children were growing up there. They would be abandoning contact with other settlements; with their legendary ancestral world, Earth; and even with the sun itself—but this can be done. It even has precedents. Throughout history, various Earthpeople have left ancestral homes to settle permanently elsewhere. (My own parents did this.)

The people of the settlement would be better off, in fact, for they would not be leaving home, they would be taking home with them.

Nor need there, in that case, be a problem as far as fuel is concerned. Since the settlers would be making no attempt to build up a huge velocity, but would be content to coast at no more than tens or hundreds of kilometers per second, little fuel would be necessary. Working their way through the comet cloud at the rim of the sun's planetary system, they might pick up a small comet or two and carry it along. Each comet would be a source of frozen volatile matter— hydrogen, oxygen, carbon, nitrogen—to serve as replacements for inevitable loss aboard ship, and as fuel for the hydrogen fusion plant.

In fact, space might not be as empty as we think, and there might be no decade that would pass without the opportunity to inspect some asteroidal object or another. Occasionally, a store of non-volatiles might show up that would give the starship an opportunity to flesh out its stores of metals and ceramics.

Eventually—a long "eventually," many generations, perhaps—the starship might approach a star. It might not be accident. Undoubtedly, the ship's astronomers would study all stars within reasonable distance, choosing one that would have a high probability of containing habitable worlds, and the starship might head for that.

There might then be a chance for a landing, for a long "stretching of the legs," for an opportunity to rebuild the starship completely, or to build another, from scratch, along improved designs. The starship might then take off with some of its people, while others might be left behind on the planet.

Those left behind could experience, for many generations, the joys of expansion and growth, and, in the end, they might build starships that would be heading out into space from a new nucleus.

There would be many starships, perhaps, that would leave the solar system—and every other colonized planet. Each planetary system would be like a dandelion gone to seed, sending new germs of life outward in all directions. The various starships, after long separation, would produce cultural and biological variations that would produce an infinite richness of experience and culture and might even produce new human species—an overflowing variety of everything that could not conceivably be duplicated on a single world or in a single planetary system.

Different cultures might have a chance to interact when the paths of two starships intersected.

Detecting each other from a long distance, the approach of two starships might be a time of great excitement on each. Each, after all, would have compiled its own records, which it would now make available to the other. There would be descriptions, by each, of sectors of space never visited by the other. New theories and novel interpretations of old ones would be expounded. Literature, music, and works of art would be exchanged, differences in customs explored.

No one starship would have time, even in millions of years, to explore more than an insignificant fraction of the vast cosmic expanse, but as each starship seeded planets here and there which, in their turn, gave rise to many new starships, all, together, would thread their way through the habitable portions of the galaxy and might even make their way out to neighboring galaxies.

And, in the process, it might well be that human beings would, sooner or later, encounter the expanding sphere of starships started by nonhuman civilizations, some of them older and more advanced than ours—from whom we can learn all the more.

It is typical of ourselves at our present primitive stage of history to think of such contacts with aliens in terms of war and conflict, but space is large enough to have room for all, and it is at least possible that curiosity may prove a stronger force than suspicion. Intelligence may even prove a strongly bonding influence so that a "cousinship of intelligence" may be recognized and differences in the physical housing of that intelligence may be dismissed as of no importance.

But what would drive starships out of the sun's planetary system in the first place? (Or out of *any* planetary system?)

We might, in lofty fashion, talk of curiosity, of the longing for adventure, and of the desire to see strange regions—but the length of the voyage and the time it would take before encountering anything of importance would be such that the generation beginning the trip could be sure of seeing nothing at all that would be interesting or curious or exciting. Why, then, should they leave the familiar regions of the solar system?

There may be compelling reasons.

Each space settlement will have its own ecological balance, which will undoubtedly be much simpler than the one on Earth. The space settlers on each little world will undoubtedly try to exclude weeds, vermin, and disease germs insofar as this is consistent with a varied and useful ecology that can maintain its own equilibrium indefinitely. It could be that each different space settlement would have a somewhat different ecosystem.

In that case, how will trade be carried on? How will human beings from different settlements work together on projects in space? Here on Earth today we have regulations forbidding the importation of plants and animals from one country into another—quarantine regulations to prevent the spread of pests or disease. How much more tightly such rules would have to be enforced in space settlements!

Some settlers might be resigned to inevitable infection and might decide that, in the end, there would be a single "space ecosystem," with minor variations from settlement to settlement. Even people who were reconciled to this, however, might not want to import parasites and diseases from Earth itself, with its vast and unruly ecosystem.

Some settlements, therefore, might well withdraw into biologic isolation. They could argue that it would be better to cut down on trade and on cooperation rather than risk the pollution of what they might consider the ideal ecosystem of their own small world.

And, in the end, in search of the ultimate quarantine, they could leave the solar system, taking themselves and their plants and animals off into the purity of distant space where nothing but the slow forces of evolution (or the faster work of deliberate genetic engineering) could change the balance they had set up.

It might be *this* which would be the motivation for the seeding of the universe—not curiosity, or the desire for excitement and adventure, but the fear of other worlds and a horror of pollution.

And this may make it possible for human beings (and their cousin intelligences, if any) to extend their horizon until, if there is time enough, it becomes coterminous with the bounds of the universe.

# PART II

## The Horizons of Time

# 13

# The Age of History

## THE CALENDAR

So far in this book, I have described how human beings have stretched out the horizons of space, little by little, to the limits of the universe. What about something that is as fundamental as space and is as constantly measured—time?

It is possible that only human beings, of all living species, do not live entirely in the present. Perhaps only human beings can remember the past in reasonable detail, pay attention to it carefully, and, out of experience of the regularity of certain changes, learn to anticipate the future in reasonable detail.

It is, of course, difficult to trace just exactly how human beings have developed their notion of time. Presumably, it began with the observation of the alternation of light and dark. From this primitive unit of time, the "day," longer and shorter periods were determined and made use of.

The day itself is established as the result of a periodic astronomic phenomenon, the endless march of the sun around the sky. This is actually the result of Earth's rotation, but it was not until after Copernicus's time in the fifteenth century that Earth's rotation began to be generally accepted as a fact, and the daily turning of the sky was understood to be an illusion.

Because of the tipping of Earth's axis, the sun also slowly moves higher in the sky from day to day, then lower, then higher, and so on, in a period that is much longer than a day. Upon this apparent up-and-down motion of the sun depend the seasons, and life on Earth pulses with the seasons. In various parts of Earth, there is a succession of wet and dry seasons, or warm and cold seasons. Plant life grows, dies, and is reborn, while animal life migrates, going and returning. Of all of this, human beings become aware.

Long before history began, therefore, units of time longer than a day have had to be worked out.

Next to the sun, the most noticeable object in the sky is the moon, and it undergoes a periodic change, almost as noticeable as day and night, in the form of its changing phases. From a thin crescent immediately after sunset (the "new moon"), it enlarges and moves farther from the sun, till it is a full moon rising at sunset; then it shrinks and moves closer to the sun until it is a thin crescent barely visible just before sunrise. Soon after this, another new moon appears in the sky just after sunset. These phases arise as a result of the moon's revolution about Earth and its changing position, therefore, relative to the sun.

The entire cycle is completed in 29.53 days, and to those who first worked out the period, this meant that the interval between new moons was sometimes twenty-nine days and sometimes thirty days. This interval was the "month," and lunar calendars have been based on this month for thousands of years. The religious calendar of the Jews and Moslems, even today, is lunar in nature.

The week is an artificial unit. It may have been meant to mark the coming of the chief phases of the moon: new, first quarter, full, and last quarter. These are 7.38 days apart. To be strictly lunar, the week would have to be sometimes seven days long and sometimes eight. The people of the Tigris-Euphrates Valley, who were the most advanced astronomers in the early days of civilization, were probably influenced, however, by the sun, the moon, Mercury, Venus, Mars, Jupiter, and Saturn, the seven heavenly bodies easily visible to the unaided eye that moved in complicated fashion against the fixed-pattern background of the stars. By granting divine honors to the sun, the moon, and the planets and allowing each one day, the notion of the invariable seven-day week arose. It was passed on to the Jews, then to the Christians, and is still with us. It is a very inconvenient unit that does not fit evenly with either the month or the year.

It is easier to mark off the cycle of the seasons by counting the months than by counting the days. While the seasons are not mathematically regular, as are the day, the week, and the month, they are a most essential cycle, whether human beings are food gatherers, hunters, or farmers. It was eventually the experience of human beings that the season cycle, or "year," which marked the period of the revolution of Earth about the sun, was a little over twelve lunar months long.

Actually, the year is 12.37 lunar months long, and the Babylonians worked out a pattern of years, sometimes twelve months long and sometimes thirteen, a pattern that repeated itself every nineteen years. This was adopted by the Greeks and the Jews (and is still used today in the Jewish religious calendar).

The Egyptians had a simpler seasonal cycle and depended almost entirely on the one annual event of the flooding of the Nile, which came, on the average, every 365 days. They set 365 days to the year, therefore, and filled it with twelve months of thirty days each, followed by five days of celebration. The months did not fit the phases of the moon, but the Egyptians didn't mind that. This was the first solar calendar.

The Romans eventually adopted the Egyptian calendar in 44 B.C. and, with the help of the Greek astronomer Sosigenes, added the additional refinement of a leap day. Since the year was actually 365 1/4 days long, every fourth year was given 366 days.

As a matter of fact, the year is not exactly 365 1/4 days either, but is 365.2422 days long—a trifle shorter. This means that three times every four centuries, a year that would ordinarily be a leap year should not be. The necessity for this was first pointed out by the English scholar Roger Bacon (1220–92), but it is very difficult to change a calendar at any time. The change was not carried through successfully until Pope Gregory XIII (1501–85) decreed it in 1582, and even then, only Catholic Europe followed him at first. Nevertheless, the new "Gregorian calendar" spread, and it is now worldwide.

There are no accepted periods of time that are longer than a year, except those that are multiples of a year. A "decade" is ten years, a "century" is 100 years, and a "millennium" is 1,000 years. These terms are from the Latin words for "ten," "hundred," and "thousand."

Less used is a "lustrum," which is five years. The word comes from the Latin for "wash" or "purify," since the Romans underwent a census every five years in their later history and the people were ritually purified thereafter, as though to make a fresh start. The Romans also had a period known as the "indiction." A new tax assessment was made at every indiction, a period of time equal to fifteen years.

Under the Greeks, an "Olympiad" was a period of four years because the Olympic games were held every four years (as they still are today in a revived cycle).

Finally, there is the "generation," which represents the average age difference between parents and children. There is no generally agreed-upon length for the generation, though sometimes it is taken to represent twenty-five years, thirty years, or thirty-three years.

## LIFETIMES

In primitive times, the average life expectancy might not have been more than thirty-three years even under favorable conditions, so that a generation was literally the time it took to replace parents by their children. Even in primitive times, however, there were people who lived longer than that.

Thus, the Bible has a famous verse which reads, "The days of our years are threescore years and ten" (Psalms 90:10). This seems to set seventy years as a normal lifetime if there is no premature cutoff through disease or violence (which, before recent times, there usually was).

There could not have been many people lucky enough to achieve that age in ancient times, but things have changed. In the world as a whole, the life expectancy is now fifty-five years, but this takes into account many regions in which modern medicine has not yet taken hold. In Afghanistan the life expectancy is 37.5 years. In Angola it is 33.5 years, and in some other parts of Africa the life expectancy for males is as low as twenty-five years.

In the United States, however, the life expectancy for white females is 77.2 and for white males 69.4. This means that more than half the white population will reach the biblical threescore and ten.

People live beyond the average life expectancy, too, of course, and did so even when that figure was very low.

Among the historical figures of the past who have lived to be over ninety are the Greek orator Isocrates (436–338 B.C.), the Roman historian Cassiodorus (490–583), the Byzantine general Narses (478–573), the Venetian doge Enrico Dandolo (1108–1205), the Italian painter Titian (1477–1576), the French scholar Bernard de Fontenelle (1657–1757); the French diplomat Louis de Richelieu (1696–1788), and the French chemist Michel E. Chevreul (1786–1889).

There is nothing astonishing in having someone attain such an age even when the life expectancy is low. There are always some people who survive the chances of disease and violence. Once such people reach the age of sixty, they have developed sufficient immunities to avoid many infectious diseases. If the person is male, he is then no longer liable to death in battle; if female, to death in childbirth. Those of the middle or upper classes are not likely to starve. In those cases, the additional life expectancy was as high in ancient and medieval times as it is now, for the killers after sixty are the degenerative diseases—cancer, atherosclerosis, arthritis, kidney disease, and the slow decay that is inseparable from age—and these are as deadly now as they were then.

Since more people reach the age of sixty now, we also have more people reaching the patriarchal nineties. There are supposed to be about 25,000 people in the world who have reached their hundredth birthday, and of these half are to be found in the United States.

There are tales of places where many people live to be far older than a hundred—in the Caucasus, in the Andes, and so on. What all these areas have in common, however, is the absence of birth certificates.

It is only recently in world history that accurate records have been kept of births and, in the absence of those records, age (which notoriously contracts among the middle-aged) tends to expand in later years out of vanity. The ages cited earlier of famous people in ancient and medieval times may have been exaggerated, though this is not necessarily so. Any age which claims to be far beyond a hundred, with no documentation to prove it, is virtually certain to be exaggerated.

As we go back into legendary times, ages are very often exaggerated far beyond all credibility. As we look through the Bible and work our way back in time, we find that Joshua died at the age of 110, Moses at 120, Jacob at 147, Isaac at 180, Terah (Isaac's grandfather) at 205, Eber (Terah's great-great-great-grandfather) at 464, Shem (Eber's great-grandfather) at 600, Noah (Shem's father) at 950, and Methuselah (Noah's grandfather) at 969.

There have been attempts at finding some natural way of explaining these great ages—as, for instance, to suppose that the very extended ones were counted in lunar months and not in years, so that Methuselah, living 969 lunar months, actually lived 78 1/2 years.

Such explanations are most unconvincing, however. It is much more likely that the ancient Jews, in mentioning these early patriarchs, were dealing with versions of even more ancient Babylonian legends, which attributed ages in the tens of thousands of years to kings reigning in legendary times. The biblical writers, rightly feeling skeptical over this, reduced the ages to mere hundreds—not enough of a reduction, of course, but at least a step in the right direction.

If we restrict ourselves to people whose birthdates have been amply documented, then the greatest human age commonly cited is that of Pierre Joubert of Quebec, Canada, who was born on July 15, 1701, and who died on November 16, 1814, at which time he was just 113 1/3 years old.

Human beings have a surprisingly long life span. If mammals are considered as a group, it turns out that the larger they are, the longer (on the average) they live. Pygmy shrews are generally dead of old age by the time they are a year old at most, even if fed well and protected from all misadventure. Tigers and lions can live into their twenties, rhinoceroses into their thirties and forties, hippopotamuses and elephants into their fifties and sixties.

The greatest age reported for a nonhuman land mammal was that of an Indian elephant which made it to sixty-nine, after having been well cared for in a zoo all its life. This means that any human being who celebrates his seventieth birthday views a world in which the only land mammals that were alive at his birth and that are still alive are other aged men and women. (It is possible that the largest whales outdo land mammals and that some may live to be ninety, but even these merely approach and do not equal the extreme human life span.)

Human beings do not share their unusual longevity even with other primates. Gibbons have not been recorded as living past thirty-two, orangutans past thirty-four, gorillas past forty, or chimpanzees past fifty-one.

The discrepancy shows up more plainly if we consider heartbeats. The smaller an animal, the more rapid its heartbeat and, so to speak, the faster it lives. If we add up the number of heartbeats that take place in the course of an animal's maximum life span, it turns out that in a surprising number of cases, the total turns out to be in the neighborhood of a billion. Whatever the size, in other words, the mammalian heart seems to be good for a billion beats and no more.

The one exception seems to be the human being. At the age of seventy, the human heart has already beat 2,500,000,000 times, and total heartbeats of 4,000,000,000 are possible in extreme cases. Why this should be is not known.

Birds generally live longer than mammals of the same size, which is puzzling, in a way, for birds have a higher body temperature and a higher metabolic rate and would therefore seem to live more quickly. On the other hand, they have a more efficient respiratory system than mammals have and that may help.

There are a great many stories of extremely old birds. Tales of birds living for more than a hundred years are common, but extended ages always tend to be exaggerated. No bird whose lifetime was followed from birth on, without possibility of error, has been known to live even as long as seventy years.

It may well be, therefore, that the human being is the longest-lived warm-blooded animal.

It might seem that the record could be broken by cold-blooded animals, whose metabolic rate is lower, who can become torpid in cold weather, and who simply live more slowly. Among the reptiles, however, snakes in general do not live past their thirties, if, indeed, they live that long. The oldest lizards and crocodiles are only in their fifties.

That leaves the chelonians (turtles and tortoises), which are the slowest-living of all the reptiles, and here there are authentic cases of individual tortoises that have lived well past the hundred-year mark. The record is thought to be for a tortoise that died in 1918 and may have been over 152 years old at the time of death.

It is possible that some fish also may live to be over a hundred years of age. This is supposedly so of sturgeons, in particular, and one sturgeon caught in Lake of the Woods on the Canadian-American border was judged to be 152 years old. There are also longevity claims for particularly large invertebrates such as giant squids and giant clams, but these are uncertain.

Plants are another thing. Many trees can live longer than any animal. The white pine can double the lifetime of the oldest tortoise or sturgeon and live more than three times as long as a human centenarian. In fact, some trees, notably the redwoods and sequoias, can live past the 1,000-year mark, and the oldest are thought to be some 4,000 years old.

The oldest of all trees and of all living things, however, is a bristlecone pine in eastern Nevada, which may be about 4,900 years old. If so, it sprouted shortly after 3,000 B.C. and is older than the Pyramids.

Suppose, though, we look at individual cells, rather than at multicellular organisms. Trees, even the largest and oldest, consist almost entirely of dead material. The living cells make up a small portion of the total mass, and none of those cells lives for more than thirty years without losing its identity through division or replacement. This is even more marked in the case of one-celled organisms, which are frequently spoken of as potentially "immortal." Perhaps so, in the sense that cellular continuity can be traced backward to the beginning of life, but this is true of larger organisms, too. The identity of a particular one-celled organism is quickly disrupted by division, sometimes after only a few minutes.

Consider the human brain cells, however. These function continuously without division or replacement for a hundred years, or a little longer, if the human being lives that long. No other cell in existence endures that long unless we consider nerve cells in a few other creatures such as the tortoise or sturgeon—and who would exchange a human nerve cell for one of those?

## CHRONOLOGY

Age, however, need not be reckoned for the individual alone. In the case of human beings, at least, speech makes it possible for old men to pass on their experiences to the young, so that a body of tradition builds up.

Unfortunately, memories are imperfect and the haze increases with the telling and retelling. Furthermore, the tales are rarely quantitative. Exact years are not remembered, and when numbers are startling they are remembered with advantage and made even more startling.

Things improved with the invention of writing, but even then records were kept skimpily and only for events that were considered of the greatest importance. What's more, prior to the invention of printing, records were difficult to make and were kept in very few copies, and were therefore easily lost or destroyed.

Prior to the nineteenth century, records were rarely kept in detail or with great attention to accuracy, and even then only in industrially advanced nations.

Even where records were carefully maintained and preserved, there was no chance of any exact "chronology"—that is, of describing exactly *when* something happened—unless there was some recognized way of identifying particular years.

The years could be identified by some locally notorious event: "the year of the big snowstorm," or "two years after the town burned down," or "three years before the dam broke." The trouble with that would be that no one outside the locality would know when the key event had taken place.

The Athenians elected a governor ("archon") each year and identified the year by the archon. In the same way, the Romans could identify a year by the consuls who were in office that year. In order for that to be of any help, however, one had to have an accurate list of archons or consuls and keep it up to date.

In the later books of the Bible, events are sometimes dated as having taken place in some particular year of the reign of such and such a king. In such a case, you would have to have a list of kings and know how long each had reigned. If every nation identified events as having taken place in a particular year of the reign of its own rulers, then one had to be able to correlate these reigns and know when King So-and-so had been crowned in terms of the reign of King Such-and-such of another kingdom. It is helpful when two kingdoms have fought a battle and each places that battle in a given year according to its own system.

On the whole, it is difficult to correlate systems, and for that reason we can rarely be sure of the exact year (much less the exact day) in which some event had taken place in early times. No one can tell, from the account in the Bible, the exact day, or, for that matter, the exact year of the birth of Jesus of Nazareth. The tradition is that he was born on December 25, and that, one week later, January 1 began the year 1 of our present system of counting. That, however, is considered to be almost certainly wrong by historians.

Apparently, the first person to initiate the simple numbering of years, without a break and without regard to kings or governors or important events, was Seleucus I, one of the Macedonian generals who succeeded to a portion of the empire of Alexander the Great after the death of that conqueror in 323 B.C.

In 312 B.C., Seleucus I captured Babylon and established himself as king over most of the Asian dominions of Alexander. He counted that as Year 1 and, thereafter had the years increase regularly in number. This was the "Seleucid Era."

The importance of the Seleucid Era was that the Jews fell under the dominion of the Seleucid monarchs in 114 S.E. (the 114th year of the Seleucid Era, which is equivalent to 198 B.C.). Thereafter, they made use of the Seleucid Era in dating their transactions, and since they carried their trading activities all over the Mediterranean world, it was possible to compare the year of the Seleucid Era with the local methods of identifying years here and there. This was an important aid to establishing a coordinated system of chronology for the entire ancient world.

Naturally, there was a convenience to setting the beginning year as far back in time as possible so that there would be less occasion to deal with events before Year 1 and thus have to deal with negative numbers. It is clumsy to have to say that Alexander the Great died in −11 S.E., for instance.

A Greek historian, Timaeus (345–250 B.C.), made this inconvenience much less likely by introducing the Olympiad system in his history of Italy and Sicily. The Olympic Games took place every four years, and Timaeus, having numbered the Olympiads, then dated events as having taken place in the first, second, third, or fourth year of such-and-such an Olympiad.

But then, when was the first Olympiad in our own system of chronology? The year of the first Olympic Games is set, by tradition, at 776 B.C., although we can't be certain that that is correct, of course.

Eratosthenes, the Greek scholar who first estimated the size of Earth, was also the first to undertake a systematic study of chronology, trying to date the list of events in different nations according to a fixed rule. He too made use of the Olympiad system. He went back even farther than 776 B.C., however, and tried to date the earliest historical event in Greek history, the Trojan War. Eratosthenes fixed the date of the fall of Troy as 408 years before the first Olympic Games, and therefore placed it in what we would call 1184 B.C.

Because of all this, Eratosthenes is considered the "father of chronology."

The Romans made their own attempt at a system of universal chronology, by numbering the years from the date of the founding of Rome. Naturally, no one knew in what year Rome had been founded, and Romans were forced to accept guesses and traditions. The Roman scholar Marcus Terentius Varro (116–27 B.C.) set the founding of Rome in the third year of the fifth Olympiad, or 753 B.C.

Thereafter, the Romans counted the years from that time. The assassination of Julius Caesar, for instance, took place in the year 709 A.U.C. ("Ab Urbe Condita," meaning "from the founding of the city"), or 44 B.C. according to our own system. Again, the fall of Constantinople in 1453, which was the final end of the Roman Empire, took place in the year 2206 A.U.C.

Naturally, the establishment of a system of chronology didn't mean that all the years assigned to particular events in that system were necessarily correct, but it remained a vast improvement over what went before.

An efficient way of counting the years is not sufficient, in itself, to place an event on the exact day. The earliest human event in history that is known to the exact day is that of a battle between the armies of Lydia and Media, and that is so only through an astronomical accident. The battle, at some spot in Asia Minor, was called off because an unexpected total eclipse of the sun frightened both armies. Astronomers can calculate backward and can determine the day on which a total eclipse of the sun could be seen from Asia Minor in the century of that battle, and we therefore know it was fought, or almost fought, on May 28, 585 B.C.

The early Christians made use of the "Roman Era" (or the "Era of Varro") in their dating, but it must have seemed wrong to them to refer all dates to the founding of a pagan city which for centuries had persecuted the Christians.

Since Christian belief was that the entire history of the world had changed with the birth of Jesus, it seemed to make sense to divide that history into two parts, the portion after the birth and the portion before it. Naturally, one had first to establish the year of Jesus' birth in terms of the Roman Era.

The first person to attempt this was a sixth-century Christian scholar, Dionysius Exiguus, who worked in Rome. About 525, he tried to work out the birth year of Jesus and decided it was December 25, 753 A.U.C.

According to this system, January 1, 754 A.U.C., is the beginning of the year we now call A.D. 1 ("anno Domini," or "the year of the Lord"), while December 31, 753 A.U.C., is the last day of the year we now call 1 B.C. ("before Christ"). This system is called the "Christian Era."

The Christian Era did not immediately catch on. The first scholar to use it in his historical writings was the Englishman Bede (673–735). Charlemagne (742–814) directed its use in his large empire, but it did not come into general use throughout Western Europe until some time after 1000. Nowadays, of course, it is in universal use, in the Christian and non-Christian world alike (at least for secular and international purposes). This is chiefly because during the period between 1750 and 1950, Europe dominated the world.

Oddly enough, Dionysius made a mistake. The Bible clearly states that Jesus was born during the reign of Herod the Great, who, it is quite certain, died in 750 A.U.C., which is 4 B.C. by the Dionysian system. Consequently, Jesus could not have been born after 4 B.C. (that is, four years before his birth). Some scholars feel he may have been born as early as 11 B.C.


## HISTORIC TIME

Given a decent chronological system, how far back can history be traced if the best possible years are attached to historical events?

In Greek history, as I said, the earliest historical event of significance was the Trojan War, which was placed at 1184 B.C. To be sure, there were earlier events that many Greeks accepted as genuine, such as the exploits of Hercules, Theseus, Perseus, and so on. Most of the later Greek scholars, however, considered them legendary and did not waste much effort on them.

There were, however, civilizations older than that of Greece. Egypt, for instance, experienced a high civilization when the Greeks were still barbarous tribesmen. An Egyptian priest, Manetho, who lived about 300 B.C., wrote a history of Egypt in Greek. He divided the rulers of Egypt, from the time when the land first became a unified nation down to his own day, into thirty "dynasties" (that is, groups of rulers of a single family). This system has been used ever since.

The third king of the 12th Dynasty was Sesostris III, and in the seventh year of his reign, it is recorded, Sirius rose at sunrise at the time of the flooding of the Nile. It is easy for astronomers to calculate when Sirius rose in this fashion, for it happens once every 1,460 years. It happened in 412 B.C., for instance, which was long after the time of the 12th Dynasty. It also happened in 1872 B.C., and that is taken as the seventh year of the reign of Sesostris III. Everything else could be calculated forward or backward from that.

Events earlier than the time of Sesostris III may be off by a decade or so, but historians seem reasonably confident that the Great Pyramid dates from about 2500 B.C., some thirteen centuries before the Trojan War, and that Egypt was first unified by Narmer of the 1st Dynasty about 2850 B.C.

Slightly older than the Egyptian civilization was that of the Tigris-Euphrates Valley, where writing was developed by the Sumerians shortly before 3000 B.C. This was the first system of writing anywhere in the world, and from that moment, history begins, since it meant that events could be recorded. They might, of course, be recorded mistakenly, or with distortions (either accidental or malicious), but it represented an advance over unwritten tradition.

History, therefore, is now 5,000 years old.

One group of early accounts of human history that is of particular importance to our own Western civilization is that which is based on the Bible. To the Jews and Christians of pre-nineteenth-century times (and to many of them even today), no part of the Bible is legendary. It is all the inspired word of God and is literally true in every part.

The Bible does not make use of a systematic chronology, which creates problems. In the later books of the Old Testament, however, there is occasional mention of monarchs that were known to the Greek historians. The Judean king Josiah, for instance, died in battle with Necho, an Egyptian pharaoh, the date of whose rule was known. It can be stated confidently, then, that Josiah died in 144 A.U.C. or 609 B.C. Using such statements in the Bible, one could then work backward, checking the calculations whenever there are clear references to events involving Assyria or Egypt.

Eventually, one could get back to 1020 B.C., as the date for the accession of Saul, the first king of Israel. This was 164 years after the traditional date of the Trojan War. This was not satisfactory, for the Jews and Christians, out of national and religious pride, wanted to emphasize the greater age of their own history as compared with that of the pagans. They therefore made an attempt to trace biblical events back to Abraham, who was viewed as having broken with the worship of idols and as having established the worship of the God of the Bible.

A church historian, Eusebius of Caesarea (260–340), reasoned out the time of the birth of Abraham from various statements in the Bible and decided that it took place in 2016 B.C., twelve and a half centuries before the founding of Rome, and eight and a third centuries before the Trojan War. This lent a satisfactory primacy of age to Judeo-Christian history and introduced the "Era of Abraham."

If, however, one calculated backward in the Bible to the birth of Abraham, then, since all the ancestors of Abraham were given, right back to the Creation, with the age at which sons were born to each, it was possible to calculate the year in which God created Earth, in accordance with the description in Genesis.

If that was done, one would have a "Mundane Era" (an "Era of the World"), one from which all things can be dated and for which no negative years are possible.

The Jews began to use such a system in the 800s, and by the 1100s it had become traditional with them, and had reached the form that still exists today. By the calculations of the rabbis, God created Earth in the year 3761 B.C., about twelve and a half centuries before the Pyramids were built. By this account, the age of Earth, at the time of this writing, is 5,742 years.

Various Christian scholars made calculations of their own and decided that the Creation took place in 5500 B.C., with an arguable difference of a decade this way or that. This would make the world nearly 7,500 years old.

To English-speaking Protestants, however, the calculation that had the most influence was that of James Ussher (1581–1656), an Anglican bishop of Irish birth. He worked out the creation of Earth as having taken place just 4,000 years before the birth of Jesus—that is, in 4004 B.C. Editions of the King James Bible ("the Authorized Version"), which is usually accepted as *the* Bible by the devout Protestants of the English-speaking world, generally have Ussher's chronological system placed in the margins or at the heads of the columns.

By Ussher's calculations, the age of Earth (and, indeed, of the universe) at the time of this writing is 5,985 years.

# 14
# The Age of the Earth

## UNIFORMITARIANISM

Prior to the eighteenth century, then, the exploration of time had seemed to set the ultimate horizon of the past, in earthly terms, at no more than a few thousand years away from the present. Is this really so, however? Can Earth indeed be no more than 6,000 years old?

There are signs that a great many interesting events have taken place on Earth, events that, one would think, would have required a long time to be carried through. The Greek philosopher Xenophanes (560–478 B.C.) was the first in our Western tradition to point this out.

He was a rationalist who was completely skeptical of the Greek tales of gods and heroes. He thought it ridiculous to imagine that divine beings had human shape and said that if oxen could carve statues they would carve gods who looked like oxen.

He noted, for instance, that there were seashells embedded in hard rock high in the mountains, and he drew the logical deduction that areas which are now high above sea level must once have been below sea level. Shellfish had grown there, their shells had been buried in mud, the mud had turned to rock, and the rock had been lifted high in the air. Since such changes were not taking place at a measurable rate in the world at the time, they either took place at a very slow pace, so that Earth was extremely old, or else there had been an enormous catastrophe in the near past.

Actually, the Greeks, like most people, had legends that spoke of catastrophes. Xenophanes would have sneered at the legends themselves, but he might have been willing to admit they were imperfect memories, dramatized and fictionalized, of actual events.

According to one such legend, Phaethon, the son of the sun god by a mortal woman, had tried to drive the chariot of the sun for a day. He failed, and the sun, veering out of its course, had nearly destroyed Earth before Zeus could slay Phaethon with a lightning bolt and restore order. Anything unusual on Earth could have been explained as the result of the events of that one horrible day.

In Christian Europe, Xenophanes' findings, and all similar observations, came to be explained by the great biblical catastrophe of the Flood.

There had indeed been a gigantic flood in the Tigris-Euphrates Valley early in its period of written history. The evidence for that flood was obtained by the English archeologist Charles Leonard Woolley (1880–1960), who in the 1920s discovered thick layers of water-deposited sand dating back to about 2800 B.C.

River valleys are subject to such floods, but this one of 2800 B.C. seems to have been unusually bad. It must have killed much of the population of the valley and may have nearly destroyed the civilization altogether. To the inhabitants of the region it might well have seemed as though the end of the world had come, and the tale undoubtedly did not suffer in the retelling. Eventually, the event was described as a worldwide flood, with only one family surviving.

The tale was told in dramatic form in the Epic of Gilgamesh, which originated with the Sumerian civilization soon after the flood, and which spread throughout the civilized world of the time. It was picked up by the Israelites, and eventually, two different versions were recorded in the Bible in Chapters 6 through 9 of the Book of Genesis. The date, as given by the Ussher chronology, was 2349 B.C.

The biblical Flood is described as having covered the entire world and as having killed every land-living animal except Noah, his wife, his three sons, his three daughters-in-law, and the various animals he had taken into the Ark with him.

Everything, therefore, that seemed awry on Earth was blamed on the Flood. All remains of ancient living things were said to be animals drowned in the Flood. All rocks where they ought not be were said to be rocks thrown about in the Flood. Ancient speculations about slow changes on Earth, about the formation of river deltas, about changes in sea level, about erosion, were all brushed aside as Flood-related. Earth was formed in 4004 B.C. (or at some date not too far removed) and it underwent this tremendous cataclysm 1.655 years later, and that was it.

And yet not everyone was satisfied; 6,000 years was not enough.

The first modern scholar to say so openly and in detail was the French naturalist Georges L. L. de Buffon (1707–88). He spent fifty years writing and publishing a huge one-man encyclopedia on natural science in thirty-six volumes (he had planned fifty). In the fifth volume, published in 1778, he gave his ideas on the development of Earth.

He felt the need for more time to account for the changes Earth had visibly undergone, and suggested that Earth had come into existence not 6,000 years before but 80,000. He suggested it was not the word of God that had done the job, as stated in Genesis, but the collision of the sun with a similarly heavy body. (Buffon called the intruding body a comet, for at the time there was no knowledge of the nature of comets.)

He then described the course of Earth's development. It began in the molten state, and slowly solidified, but not before rotation had ballooned its equatorial regions into the bulge actually observed. The solid crust of Earth then wrinkled into mountains, and the water vapor in the atmosphere condensed and formed the oceans. Sea life formed about 40,000 years ago, and some of the early organisms were trapped in mud so that they formed fossils. Some of the water then drained away, through cracks, into Earth's interior, so that dry land was exposed. On the dry land, animals appeared and, finally, man. Buffon even imagined a form of evolution through degeneration—some horses degenerating into donkeys, some men into apes, and so on.

Although Buffon reasoned things out carefully and made use of observation and, when he could, experimentation, he could not stand up against the force of offended religion. He was forced to retract his views.

In 1789, however, the year after Buffon's death, the French Revolution was initiated and the power of the churches was greatly weakened. It became possible for scientists to ignore the ancient Babylonian legends, even in their biblical form, and to consider the history of Earth on the basis of evidence. The horizon of time began to stretch backward.

The task of stretching it was taken up by a Scottish geologist, James Hutton (1726-97). His careful studies of Earth's terrain convinced him—as it had convinced others before him—that there was a slow evolution of the surface structure. Some rocks, it seemed clear to him, were laid down as sediment and compressed; other rocks were molten in Earth's interior and were then brought to the surface by volcanic action; exposed rocks were worn down by wind and water; and so on.

Hutton's great addition to all this was the suggestion that the forces now slowly operating to change Earth's surface had been operating in the same way and at the same rate through all Earth's past. This is the "uniformitarian principle." He felt that the chief agent at work, driving all these slow changes, was the internal heat of Earth. The planet, in short, was a gigantic heat engine.

To Hutton, it seemed as though Earth's history must be extremely long, since although the actions involved were creepingly slow, vast changes had nevertheless had time to take place. There was, he said, "no vestige of a beginning, no prospect of an end."

In 1785, Hutton published *Theory of the Earth,* in which he advanced his notions. In so doing, he became "the father of geology."

There was, of course, strong resistance to Hutton's notions, but the religious establishment was not strong enough in Great Britain to force him to retract. Besides, even his opponents were forced to accept a long-lived Earth. What those opponents had to do was to combine the clear evidence of a long-lived Earth with the short-lived Earth apparently described by the Bible.

The Swiss naturalist Charles Bonnet (1720–93) tried to explain the increasingly numerous fossils being discovered in the rocks by supposing that every once in a while, Earth underwent a catastrophe that killed all life upon it, leaving traces behind in fossil form—after which everything started anew. This was "catastrophism" as opposed to uniformitarianism.

Catastrophism made it possible to state that even though Earth had existed for a long time, people might suppose that there was one last catastrophe 6,000 years ago, after which God created Earth as it now exists and as it is described in the Bible.

Catastrophes allowed scholars to avoid assuming that life forms evolve and change. Although the fossils and their resemblances, with crucial differences, to living organisms made the thought of evolution inevitable, that could be denied if it was firmly held that fossils belonged to past creations and had nothing to do with present life.

Another Scottish geologist, Charles Lyell (1797–1875), took up Hutton's views and his uniformitarianism. Between 1830 and 1833 he published *The Principles of Geology* in three volumes. The book was well written and sold well. The careful presentation and analysis of the evidence helped establish uniformitarianism,* and the question of whether Earth had lasted more than 6,000 years was no longer a matter of controversy among scientists. The answer was that

---

* Since Lyell's time. extreme uniformitarianism has been abandoned by geologists. While the course of events is slow and steady for the most part. there *have* been occasional violent events. These have never managed to wipe out life altogether. but one or two seem indeed to have done much damage. No such events have taken place in historic times. however. In other words. there was no worldwide Flood. as described in the Bible. in the third millennium B.C.

it had indeed lasted more than 6,000 years, and even much, much longer.


## THE CONSERVATION OF ENERGY

The question was: How much longer?

If one accepts uniformitarianism, one might measure the rate at which some continuous change is taking place and then calculate how long it would take to produce some observed end result. How quickly is the Nile Delta growing and how long did it therefore take for it to form? How rapidly are sediments being laid down and how long did it therefore take for an observed thickness of sedimentary rock to form?

Or suppose the ocean, to begin with, was fresh water and that it gradually grew salty through the traces of salt leached out of the land by the many rivers. If the salt content of river water is analyzed and the total river water draining into the ocean is estimated, it was possible to argue that it took 1,000,000,000 years for the oceans to grow as salty as they now are.

This was not a hard and fast figure, for one could not be sure that the rivers had behaved in the past exactly as they do now, or that some shallow arms of the sea had not been pinched off and evaporated, leaving salt behind on dry land and subtracting that from the oceans.

Nevertheless, all the methods of estimating Earth's age by measuring the slow rates of geologic processes and considering the large results made it clear that Earth was at least hundreds of millions of years old.

This was convenient for geologists, of course, but it was also convenient for biologists, thanks to new considerations of the slow development of life forms.

An English geologist, William Smith (1769-1839), had carefully studied the fossils and showed that various layers of rock ("strata") had fossils characteristic of themselves; that one could follow the strata across kilometers of countryside and always find those characteristic fossils in each stratum. What's more, the deeper the strata and, therefore, presumably the older they were, the more different the fossils were from living organisms. Smith published his observations in 1816, and this gave a great push forward to the notion of biologic evolution—the slow change of organisms from species to species, from early forms to present-day forms.

Of course, it was not so much the fact of evolution that bothered biologists but the mechanism. What was it that *caused* species to undergo changes?

In 1809, the first attempt to answer that question was made by the French naturalist Jean Baptiste de Lamarck (1744–1829). He suggested that animals changed somewhat in the course of their lifetime as a result of their natural activity. Thus, an antelope feeding on leaves would constantly stretch upward for them and lengthen its neck slightly in the process. This longer neck would be passed on to its young, which would continue the process until in the end the antelope would become a giraffe. This "inheritance of acquired characteristics" did not explain enough, however. It wouldn't account for how the giraffe gained its splotched coat, something that is not characteristic of other antelopes. The splotched coat was useful as natural camouflage, but did giraffes *do* anything that tended to make the coat splotchier?

In 1859, however, the English naturalist Charles Robert Darwin (1809–82) published his *Origin of Species,* in which he advanced the notion of "evolution by natural selection." In every generation, there would be random variations from individual to individual, and those individuals who were faster, or stronger, or shrewder, or had more endurance, or possessed a coat that blended better with the environment, had a better chance to live and reproduce. In the long run, there would be progressive changes that would better the adaptation of a species to its environment. Enough change would produce a new species that would be an improvement on the old in certain ways.

This notion met with almost hysterical resistance from the conventional religious leaders of the day (and still does now), but it was accepted by the scientific community and by those who reasoned from a basis other than tradition, authority, and emotion.

What was needed, however, if Darwinian evolution was to work (or any of the modified and improved variations thereof that have been advanced since), was a great deal of time. Unguided natural selection, choosing among small random changes, most of which are for the worse, takes a long time, and many millions of years are clearly needed.

It was therefore upsetting when, even as Darwin was advancing his theory, the lifetime of Earth seemed suddenly to contract. It didn't shorten to 6,000 years, but it did shorten by an uncomfortable amount as a result of new findings in physics.

In the 1840s, it was becoming more and more obvious that there were many forms of energy, and that one form could easily be transformed into another. If, however, heat was taken into account as one form of energy, the transformations would be carried through without net gain or loss.

The German physicist Hermann Ludwig von Helmholtz (1821-94) finally put it into words, clearly and convincingly, in 1847, and with that the "law of conservation of energy" was established. It stated that although energy could be changed from one form to another, it could be neither created nor destroyed.

Having established that principle, Helmholtz began to consider the sun.

It had always been assumed that the sun was eternal or that it would exist until such time as it pleased God to put it out. It was viewed as a container of light and that light illuminated Earth.

Yet light sources here on Earth only continue to do the work of illumination as long as there is something to burn. When the fuel is gone, the light goes out. One might argue that heavenly light follows other rules, but Helmholtz felt that the law of conservation of energy had to be universal. The sun poured out enormous quantities of energy, and that had to come from somewhere.

If the sun were a huge coal fire, then at the rate that it was emitting energy, the whole substance of the sun would have burned up in 1,500 years. Since the sun had clearly been in existence for at least 6,000 years, even according to the strictest believers in the Bible, there had to be some other and more copious source than burning coal.

Helmholtz considered several possibilities and found each of them wanting. His final conclusion, in 1854, was that the sun was contracting. Its substance was falling inward in response to the pull of gravity, and, as it did so, the energy of the motion of falling ("kinetic energy") was converted into radiation. So vast was the sun that even a small degree of falling supplied all that it needed for its radiation. Throughout the history of civilization, the total contraction of the sun was too small to be detectable by the instruments of Helmholtz's day.

But what if Earth was vastly older than civilization was? If we imagined ourselves reaching farther and farther back in time, the sun would have had to be larger and larger. It would not take many millions of years into the past before the sun would have been large enough to engulf Earth, and Earth could not be older than that as an absolute maximum.

It didn't seem likely that this maximum was going to be long enough for the geologists to account for observable changes in Earth's crust, or for biologists to account for evolutionary changes.

This was made quite plain by the work of the Scottish physicist William Thomson, later Lord Kelvin (1824-1907). As early as 1846, he showed that if Earth had originated as part of the sun (as Buffon had maintained sixty years before), then it would have cooled down from the sun's surface temperature to its own present surface temperature in perhaps 100,000,000 years.

Then, too, Earth had an equatorial bulge produced by its rotation, but its present rotational speed is not quite high enough to produce such a bulge. However, Earth's rotation is gradually slowing because of the moon's tidal action. There must have been a time in the past when the rotation was fast enough to produce the bulge, and at that time Earth must have been hot enough to be liquid. After the bulge was produced, Earth cooled to the point of solidification and the bulge was frozen in place. The bulge remained as it was as Earth's rotation continued to slow and became too slow to account for the size of the bulge. Kelvin's calculations showed that on this basis, too, Earth had solidified and become a possible abode for life not more than 100,000,000 years ago.

This Kelvin had done before Helmholtz had published his theory about the sun's maintaining its radiation through contraction. Once that publication appeared, Kelvin calculated how much radiation would be produced if the sun began as a globe 300,000,000 kilometers in diameter (and therefore filling Earth's orbit) and had then contracted to its present size. At the present rate of energy release, the sun would have had to carry through that contraction in 25,000,-000 years. Earth obviously could not have existed when the sun was so big that its surface extended to Earth itself.

It seemed clear by these various lines of argument, then, that Earth could not possibly be more than 100,000,000 years old and that, very likely, it was not more than 25,000,000 years old. And this was simply not long enough, by a good margin, for geologists and biologists.


## RADIOACTIVITY

But then, in 1896, the French physicist Antoine Henri Becquerel (1852-1908) discovered, more or less by accident, that certain substances emitted radiation continually. The phenomenon was called "radioactivity." Continued investigation of these radiations by many physicists showed that it was the heavy metals, uranium and thorium, that gave off the radiations, and that this came about because atoms * were breaking down. In the process, heat was continually given off. Somewhere in the heart of the atom there was an energy source, one whose existence was hitherto unexpected. This new energy source came to be called "nuclear energy."

This changed everything. To be sure, the breakdown of a single radioactive atom produced an infinitesimal quantity of heat, and very few atoms in any ordinary sample of matter were breaking down—that was why nuclear energy had gone unnoticed through the long ages of human history. Nevertheless, if all the radioactive atoms breaking down in Earth itself each second were considered, enough heat would be supplied to keep Earth at its present temperature for a long, long time.

Therefore, any calculations that considered Earth to be cooling down at a steady and rather rapid rate, from the time it was part of the sun, were invalid. Thanks to radioactivity, Earth cooled down very slowly indeed, and short life spans based on cooling rates could be thrown out.

Then, too, it might be that nuclear energy, not contraction, was the source of the sun's energy and radiation. In that case, the sun might have been radiating at its present level for billions of years without any noticeable change in size, and any calculations involving its steady contraction could be thrown out too.

With the beginning of the twentieth century, then, after fifty years of agonizing doubt, it became once more possible to think of Earth as being old enough to allow geologists and biologists room for their observations and theories.

In fact, the phenomenon of radioactivity offered scientists a way of measuring the age of Earth that was better than anything ever possible before.

The New Zealand–born physicist Ernest Rutherford (1871–1937) showed, in 1904, that particular varieties of radioactive atoms broke down at fixed rates. Any single atom might break down at an unpredicted moment, but a very large number of atoms of a particular variety, taken together, followed the rules worked out for what is called a "first-order reaction."

In such a reaction, half the atoms would have broken down after a certain interval of time, say $x$ years. Half of what remained would break down after an additional $x$ years; half of what then remained after yet another $x$ years; and so on. That period of $x$ years Rutherford called the "half-life." Each variety of radioactive atom had its characteristic half-life, from very short to very long.

* Atoms are the submicroscopic units of which matter is composed. We'll have more to say about them later in the book.

The first element found to be radioactive, uranium, exists in two forms (or "isotopes") in nature: uranium-238 and uranium-235. The first has a half-life of 4,500,000,000 years and the second has one of 700,000,000 years. Since it is becoming customary to set 1,000,000,-000 years equal to "one eon," we might say that the half-life of uranium-238 is 4.5 eons and that of uranium-235 is 0.7 eons.

The second element found to be radioactive, thorium, consists of only one form, thorium-232, and that has a half-life of 13.9 eons.

In that same year of 1904, the American physicist Bertram Borden Boltwood (1870–1927) produced the final bit of evidence to the effect that uranium and thorium atoms, in breaking down, became atoms of other elements that in their turn broke down through a long chain of such events, until they ended as stable isotopes of lead. Uranium-238 ended as lead-206, uranium-235 ended as lead-207, and thorium-232 ended as lead-208.

Thinking about it further, Boltwood pointed out, in 1907, that it might well be possible to use such breakdowns to determine the age of a rock—or at least the length of time during which the rock had remained solid and undisturbed. By measuring the quantity of uranium-238 and lead-206 in a particular rock, one could calculate how long it had taken that much lead-206 to form out of that much uranium-238, knowing the half-life of the latter.

There is a catch. Not all lead is the product of uranium and thorium. Some might have been in the rock from the beginning. After all, there are rocks that contain no uranium or thorium at all, and probably never did, and yet lead may be there.

Fortunately, there's a way of allowing for that. Lead consists, in nature, of four isotopes. Three of these can be formed from uranium and thorium, but the fourth, lead-204, cannot be. Any lead-204 that exists has always existed since Earth was first formed.

In samples of lead that, as far as we can tell, have never been associated with uranium or thorium, there are always forty-seven atoms of lead-206 for every three atoms of lead-204. Therefore, if uranium and lead both exist in a rock, one must first analyze for lead-204. If that quantity is multiplied by 47/3, then one obtains the quantity of lead-206 that was present without reference to uranium. Any quantity of lead-206 present *above* that figure was produced by the breakdown of uranium.

This sort of thing doesn't give the age of Earth at once. Not all rocks contain measurable quantities of uranium. Those that do may well have experienced melting at some time in the past. Volcanic action is always melting rocks, and there may have been many periods of active vulcanism in Earth's history. Most rocks have at some time or another been melted, or broken up by erosion or solution, and, in these cases, the uranium and lead would surely have been separated, since the two elements would respond differently to geologic changes when they are not frozen into solid rock.

In order to draw a useful conclusion from the uranium and lead in a rock, then, that rock must have been undisturbed by melting, erosion, or solution for a long time. One can therefore only determine the age during which the rock has lain undisturbed, and this may be considerably less than the age of Earth.

Even so, it was possible to find rocks which, from the uranium and lead content, were clearly over an eon old. To this date, the oldest rock measured has been one from West Greenland, which turned out to have an age of about 3.7 eons, or 3,700,000,000 years.

This gives only a minimum age for Earth. It could be much older than that, but it might be that no piece of available rock has survived untouched for longer. Is there any way we can push the horizon of time beyond the age of the undisturbed rocks?

## THE FORMATION OF THE SOLAR SYSTEM

According to the biblical view, not only Earth, but the entire universe—the moon, the sun, the planets, and the stars—was formed 6,000 years ago. The time might be wrong, but if the entire universe *were* formed at the same time, then if we could determine the true age of any astronomic body, we would also have the age of Earth.

But was all the universe formed at once? Not necessarily. Buffon, for instance, had suggested that Earth was formed from the sun, by a catastrophic collision of the sun with another body. If that were so, then the sun might have been existing for countless eons before the formation of Earth. In that case, though, it might be that the various planets, at least, had been formed simultaneously.

In 1796, however, the French astronomer Pierre Simon de Laplace (1749–1827) suggested that the entire solar system, including the sun, had originated out of a vast swirling cloud of dust and gas called a "nebula" (from the Latin word for "cloud").

The cloud contracted under the pull of its own gravitation, and as it did so, its rate of rotation increased according to the "law of conservation of angular momentum." Finally, the cloud spun so rapidly and developed so great an equatorial bulge that a ring of matter broke loose from its equator altogether and condensed into a planet. This happened over and over again, until what was left of the cloud became the sun.

This "nebular hypothesis" was very popular throughout the nineteenth century. If it is accepted, then it would appear that the outermost planet is oldest and that the rest are increasingly younger the nearer they are to the sun—the sun itself being the youngest of all. Even in this case, though, the moon might be of the same age as Earth, since it is at least likely that the moon formed out of the same ring of matter Earth formed out of.

The nebular hypothesis did not retain its popularity, however. It turned out that the planets have 98 percent of all the angular momentum (that is, of all the turn-content, so to speak) in the solar system and the sun has only 2 percent. If the planets formed from rings of matter breaking away from the main mass of the contracting nebula, how could those relatively tiny rings collect all the angular momentum?

In 1900, therefore, the American geologist Thomas Chrowder Chamberlin (1843-1928) revived Buffon's notion. He suggested that another star had passed close to the sun in the distant past. The gravitational influence of each on the other tore matter out of the two. The extracted matter, pulled out in a long string, was given "English" by the gravitational pull of the stars as they passed each other. They were set to turning, and angular momentum was concentrated in them by that gravitational influence.

By Chamberlin's theory, all the planets were of equal age.

For about thirty years, this theory was popular, but closer investigation showed that there were insuperable difficulties. The gravitational pull couldn't yank out enough matter, couldn't send it out to large enough distances, and couldn't put enough twist on it. Besides, the interior of the sun was far hotter than was thought in 1900. and any matter pulled out of the sun would not cool down into planets; it would just expand into space as thin gas.

Finally, in 1944, the German astronomer Carl Friedrich von Weizsacker (1912-    ) returned to the nebular hypothesis. By then, much more was known than in Laplace's time. More was understood about how a nebula would turn, and how the outer parts would set up whirlpools and eddies. Weizsacker showed that the sun and all the planets—the whole solar system—would form more or less simultaneously. As for the angular momentum, it could now be shown that the sun had an electromagnetic field and that as the sun condensed from the nebula, this electromagnetic field could serve to transfer angular momentum to the planets.

Astronomers now feel confident that the entire solar system was formed at the same time and that the age of any part of it would give us the age of Earth. Therefore, the analysis of moon rocks, for instance, would be as useful as the analysis of Earth rocks.

In 1969, astronauts finally reached the moon, and moon rocks were brought back to Earth. It was found that the moon rocks brought back from the lunar highlands were 4.0 to 4.2 eons old, up to half an eon older than the oldest Earth rocks tested.

Still, that, too, only gives a minimum age.

As the planets formed, according to the present view of the origin of the solar system, the dust and gas of the nebula collected into fragments, which in turn collected into larger fragments, into pebbles, boulders, mountains. The cores of large bodies formed, and these swept up the debris. Toward the end of planetary formation, the debris that was left was large enough to gouge holes into the crusts of the planets, as these performed the final "sweep-up."

The last bodies that were swept up left craters that are still visible, not only on the moon, but on Mercury, Mars, and on various satellites of Mars, Jupiter, and Saturn. (Earth would have them, too, but the action of air, water, and life on our planet has eroded them away except for some faint traces of very recent meteorite strikes.)

The crashing of large bodies into the moon disturbed, powdered, and probably temporarily melted every part of the visible surface of the moon at one time or another. What we now see of the lunar surface has remained relatively quiet and unchanged only since the cessation of the bombardment of the moon. Clearly, that bombardment ceased a little over four eons ago, but how long had it lasted before then?

What we should do would be to analyze the structure of bodies still smaller than the moon. The smaller the body, the less likely it would have been for it to attract debris in the early days of the solar system and the longer it would have remained undisturbed.

It is not easy for us to reach such small bodies for testing—with one exception. There are meteors that occasionally enter Earth's atmosphere and streak downward. Some are large enough to survive the passage and strike the ground as meteorites, and some of these can be recovered and studied.

The age of meteorites has been determined by following the slow radioactive breakdown of a particular rubidium isotope into strontium, and a particular rhenium isotope into osmium by changes wrought by cosmic ray bombardment, and so on. The various methods all yield just about the same result, and the ages of the meteorites fall between 4.4 and 4.6 eons.

The conclusion, then, is that the entire solar system, including Earth, was formed about 4.6 eons ago—that is, 4,600,000,000 years ago. For about 500,000,000 years afterward, there was a gradually decreasing bombardment of the larger bodies by debris. Since a little over 4,000,000,000 years ago, the worlds of the solar system, including Earth, have been reasonably quiet. It is now thought that simple life, in the form of primitive cells, may have developed on Earth as long as 3,500,000,000 years ago, and that despite occasional periods of near-catastrophe, life has never been completely wiped out since, and has never had to start over again.

In two centuries, then, scientists have pushed back the horizon of time from the biblical 6,000 years to 4,600,000,000 years for the age of Earth—a 760,000-fold increase.

# 15
# All of Time

## THE EXPANDING UNIVERSE

But this, so far, is only a matter of the age of the solar system, which is but a mere dot in the broadness of the universe. Were all the stars in the universe formed at the same time the sun was? Is the entire universe 4.6 eons old?

It is conceivable that this is so. It is also conceivable that the universe has existed for an indefinite number of eons and that the sun and its family of planets are but latecomers. After all, in the Orion nebula, for instance, there are condensations that are even now sparking into brilliant life as newborn stars. And if one star is born now and another 4.6 eons ago, why not still another 46 eons ago?

It was hard enough to determine the age of Earth, so how can it be possible to determine the age of the universe? Yet in a way, the age of the universe is the simpler problem of the two.

The key to the determination came as long ago as 1842, when an Austrian physicist, Christian Johann Doppler (1803–53), showed that sound was lowered in pitch when the source was receding relative to the observer, and was raised in pitch when the source was approaching relative to the observer.

It was easy to interpret this where sound waves were concerned. When the source receded, the waves were stretched longer, and long waves have a lower pitch than short ones. When the source approached, the waves were squeezed shorter. This is called the "Doppler effect."

In 1848, the French physicist Armand H. L. Fizeau (1819–96)

showed that this applied to light, which is also a wave phenomenon. If a light source is receding from the observer, the light it emits is shifted toward the long-wave red; if it is approaching, the light is shifted toward the short-wave violet. Fizeau pointed out that one could follow this shift by comparing the position of the lines in the spectrum of a moving source with those in similar light from a stationary source.

Since the Doppler effect worked at any distance, it should work with a star. In 1868, the English astronomer William Huggins (1824–1910) showed there was a distinct, if small, red shift in the spectral lines of the star Sirius. In this way, he was able to demonstrate the speed at which Sirius was receding from us.

Thereafter many stars were tested in this way and it was found that some were receding, and some approaching, at speeds varying from very low ones to 100 kilometers per second or so.

Eventually, as telescopes and spectroscopes improved, the spectra of other galaxies were taken and the spectral lines studied. The first to do this was the American astronomer Vesto Melvin Slipher (1875–1969). In 1912, he determined that the Andromeda galaxy was approaching the solar system at some 200 kilometers per second. He then went on to study the spectra of other galaxies (and this at a time when they were not yet understood to be galaxies). He found, to his surprise, that all but two of these was receding from us at speeds considerably higher than the speeds of stars.

The work was continued by the American astronomers Edwin Powell Hubble (1889–1953) and Milton La Salle Humason (1891–1972). They continued to study the spectra of galaxies and found that without further exception, they were receding. What's more, the dimmer they were, the faster they were receding.

Hubble did what he could to estimate the distances of these galaxies, and by 1929 he was confident enough of the results to announce what was called "Hubble's law"—that the rate of recession was proportional to the distance of a galaxy from us. In other words, if galaxy A was five times as far from us as galaxy B was, then galaxy A was receding from us at five times the velocity. What's more, as one penetrated outward, recession speeds of thousands and tens of thousands of kilometers per second were recorded.

It seemed peculiar that there should be something about our own galaxy that was pushing all other galaxies outward, and pushing the far-distant ones more effectively than the nearer ones. In fact, it made no sense. Whatever was happening, it could have nothing to do with us.

The conclusion reached at once was that the universe was expanding; that all the galaxies were moving away from each other. (Actually, galaxies are arranged in clusters that hold together by gravitational attraction, and it is the clusters that move away from each other. The reason there are two nearby galaxies that approach us is that they are part of our own local group of galaxies.)

Therefore, if we were standing on a planet in *any* galaxy at all, all the galactic clusters outside our own would seem to be receding, and the farther away they were, the faster they would be receding. There is nothing special about *us*.

What made the notion of the expanding universe even more attractive was that Einstein's General Theory of Relativity, first presented in 1916, predicted that such an expansion would take place.

An expanding universe at once offered a way of determining how old the universe might be.

Suppose we looked backward in time and considered what happened to the universe. It would be like running a movie film backward. If the universe is expanding and growing larger as we move forward in time, then it would be contracting and growing smaller as we move backward in time.

As far as expansion is concerned, we can imagine that it could continue forever, since, as far as we know, there is no barrier out there to stop the expansion no matter how far it goes. Contraction is another thing. If the universe contracts far enough, it will shrink to zero volume, and it can go no further than that. That will have to be zero-time, the moment at which the universe began.

This sort of thing was considered by a Belgian astronomer, Georges Lemaître (1894–1966), even before Hubble announced his law. In 1927, Lemaître suggested that in the far-distant past the universe had all its mass—all its galaxies—compressed into a small volume; he spoke of that small volume as the "cosmic egg." It had exploded in an enormous cataclysmic outburst that created the universe as we know it. The outward-hurling galaxies still mark the force of that initial explosion. The Russian-American physicist George Gamow (1904–68), who agreed with this, called the explosion of the cosmic egg "the big bang," and the phrase caught on.

Not everyone accepted this notion. In the 1940s, however, Gamow pointed out that at the start, the universe must have been tiny and unimaginably hot. The radiation it emitted would be extremely short-wave and energetic. As the universe expanded and cooled, the radiation would grow steadily longer-wave. By the present time, the universe would be so large and so cool that the radiation would be of microwave length. Consequently, astronomers should detect a dim background of microwave radiation in all directions equally.

In the 1940s, when Gamow announced this, astronomers had no instruments capable of making such an observation—but the techniques of radio astronomy rapidly developed. In 1965, two American physicists, Arno A. Penzias (1933–    ) and Robert W. Wilson (1936–    ), detected the background radiation of microwaves and noted it to be coming from all directions. Since then, the notion of the big bang has been generally accepted by astronomers.

The question is: When did the big bang take place? If we mark that as the origin of the universe, the answer to the question will give us the age of the universe.

Suppose that the velocity of recession of a galaxy increases by 1 kilometer per second for every 6,500 additional light-years it is distant. That means a galaxy that is 6,500,000 light-years away from us should be receding at 1,000 kilometers per second. How long ago, if we look backward in time, was it right on top of us?

If that galaxy had traveled from a point near ourselves to a point 6,500,000 light-years away at 1,000 kilometers per second, it would take it 2,000,000,000 years to cover the distance.

What's more, a galaxy twice as far from us would be receding twice as fast, and it would have taken 2,000,000,000 years for it to cover the doubled distance at its doubled speed. An object ten times as far away would be receding ten times as fast, and it, too, would take 2,000,000,000 years.

In other words, if the galaxies are all receding from us at a rate that increases by 1 kilometer per second for every 6,500 additional light-years, then *all* the galaxies, if we look backward in time, would have shrunk together into a cosmic egg 2,000,000,000 years ago. The big bang would have taken place two eons ago and the universe would be two eons old.

Hubble's first figures actually gave this result, and once again the geologists were faced with an impossibility. Fifty years before, they couldn't accept Kelvin's figure of 25,000,000 years for the age of Earth, because they knew Earth was older than that. Now, in the 1930s, they couldn't accept Hubble's figure of 2,000,000,000 years because, thanks to the radioactive clock, they knew Earth was considerably older than *that,* and Earth could not be older than the universe.

One of the best methods Hubble had for determining the distance of the nearer galaxies was to make use of certain variable stars called "Cepheids."

Back in 1912, the American astronomer Henrietta Swan Leavitt (1868–1921) had shown that the longer the period of variation, the more luminous the star, according to a certain formula. If Hubble noted that a Cepheid could be made out in a particular galaxy, then he knew from its period how luminous it was. From its luminosity he could calculate how bright it would appear to be at any distance. From its actual appearance of brightness he could then calculate its distance.

In 1942, however, the German-American astronomer Walter Baade (1893-1960) took advantage of the wartime blackout to study the Andromeda galaxy in detail. He discovered that there were different varieties of Cepheids. Leavitt's formula applied to one variety, but not to the others, and it was the others that Hubble had used to determine distances.

New formulas were worked out for the other varieties of Cepheids, and it turned out that the galaxies were about ten times as far away as Hubble thought they were. Their speeds of recession, however, were the same as had been thought before. That speed was measured directly by the spectroscope and it didn't depend on the Cepheids, or on their varieties.

It turned out that at the same rate of recession it would take ten times as long to cover ten times the distance for the galaxies in each galactic cluster. It would take such a universe not 2,000,000,000 years to contract to a point, but 20,000,000,000 years. That would mean that the big bang took place not two eons ago but twenty eons ago.

That would be so, however, only if the rate of recession stayed the same no matter how far backward in time we went. This is not likely to be true.

Once the big bang took place, the various fragments of the cosmic egg burst outward against the pull of the gravitation of all its parts. As it moved outward, the rate of recession had to slow down, thanks to the steady pull of the universe's overall gravitational field. Right now the universe must be expanding at a rate slower than ever before.

Since the universe is expanding more and more slowly, then, if we look back in time, we would see the universe contracting faster and faster—as it should, for the smaller it gets the more powerful its own gravitational pull becomes. This means that, since it speeds up, it comes together in less than twenty eons.

A reasonable estimate, then, allowing for the effect of gravity, is that the big bang took place fifteen eons ago, so that the universe is 15,000,000,000 years old.*

It follows, then, that the universe was a little over ten eons old

---

* There is some argument about this, and recently some astronomers have presented evidence that favors a nine-eon age, but there have been some astronomers who have held out for longer than fifteen eons, too. Until further notice, it might be wise to stick to fifteen eons.

when the solar system was formed, and that the universe lived out two-thirds of its present existence without Earth.

If Earth is 760,000 times as old as James Ussher's Bible-based guess had it, the universe as a whole is 2,500,000 times as old.


## THE DEATH OF THE SUN

It would seem that we can go no further in the direction of time past, for nothing can lie beyond the beginning of everything we can observe. What about the other direction, though? What about the horizon of time future?

It would seem that there, everything is permanently obscured from us. We can remember the past, study the records of it as preserved by human beings who lived before us, deduce the past from what we can see and observe in the present—but what on Earth can we do about the future?

Actually, while we don't have a crystal ball and can't foresee with certainty the precise winner of tomorrow's horse race, or exactly which of us will be hit by a car or will meet an attractive member of the opposite sex within a given time, we can make some overall predictions.

For instance, from all the statistics we have of human births, deaths, and lifetimes, we can predict that of 100,000 sixty-year-old human beings, a certain number will be dead within the year, and a certain number will be alive five years later—and we are likely to be found to be fairly correct as time passes. We can't say exactly which ones will die and which will survive, but the statistical prediction has its uses, too.

Can we look at Earth and make estimates as to when it will die, or at least when it will cease to be an inhabitable world?

To be sure, there are chances of cataclysms which would disrupt the orderly progression of events and make prediction more difficult, but even cataclysms can be allowed for, after a fashion.

For instance, there are a number of mountain-sized objects whizzing about the sun in orbits that carry them within a few million kilometers of Earth's orbit, and every once in a while those objects and Earth are at points in their respective orbits that place them moderately near each other. Such objects are called "Earth-grazers," and in 1937, one of them, about a kilometer across in size, passed within about 400,000 kilometers of Earth.

If their orbits were permanent, Earth-grazers would never touch us, but these objects are forever being slightly affected by Earth's gravitational pull, and by the pull of other planets they approach, and so their orbits are continually being modified, and it may be that someday the orbit of some Earth-grazer may intersect Earth's orbit, and both object and Earth may eventually reach the crossing point at the same time. Astronomers can calculate that such an incident should take place once every so many million years, on the average.

Apparently, one such collision did take place about 65,000,000 years ago and (it was suggested in 1980) sent up such a cloud of dust as to obscure the sun to a large extent for three years. As a result, most vegetation died, and so did most animals. All the dinosaurs, and a number of other kinds of animals, were wiped out, and Earth was nearly sterilized. (It was the closest thing we know of to the kind of "catastrophes" suggested by Bonnet two centuries ago.) Some forms of life did survive the catastrophe, and these were ancestral to all the forms living today, including ourselves, but it was a near thing—and it could happen again.

Such cataclysms, however, are chancy things. Something quite terrible could happen next year, or not for a million years, or (just possibly) never. Are there any *inevitable* events that would mark the end of the Earth, just as the events of 4.6 eons ago marked its beginning?

If only Earth were involved, probably not. If we could imagine that Earth's surroundings stayed exactly as they are now, benign enough to support a planetful of life, then Earth might endure as it is now, forever.

But will Earth's surroundings remain benign? The most important influence on Earth, by far, is the sun. It is the sun's radiation that warms Earth and supplies the energy for air and water movements, the energy that plant life turns into tissue, thus supplying food for the animal world, and so on. It was because the asteroid collision of 65,000,000 years ago blocked much of the sunlight for three years that Earth was nearly sterilized.

Can we rely on the sun to last forever?

Certainly not!

Earth itself is in a state of equilibrium. There are crustal shifts, volcanoes, earthquakes, mountain-building, erosion, currents in its molten core and magnetic fields in near space, but these are *not* progressive things. These changes do not move in one direction only, but wobble about some average.

The sun is fundamentally different. It is losing energy at an enormous rate, pouring it out into space without chance of recovery, moment after moment, year after year, eon after eon. As Helmholtz was the first to realize, the energy has to come from somewhere, and wherever it comes from, the supply will have to be totally consumed someday.

If the sun were burning coal, and were starting fresh right now, it would be dead ash in 1,500 years if it delivered energy constantly at its present level.

If it were gaining energy by contracting, at its present level of energy output it would take only 9,250,000 years to shrink from its present size to nothing.

In either case, if the sun had started radiating at the moment of its formation 4.6 eons ago, it would be dead and gone almost that long a time ago.

Of course, it is nuclear energy that keeps the sun going, and that can last far, far longer than any other source of energy—but even "far, far longer" isn't forever. How long would the sun's nuclear energy supply last?

That depends on exactly what the nature of the nuclear energy is, of course. The basic details were worked out by the German-American physicist Hans Albrecht Bethe (1906–    ) in 1938.

In the solar core, the temperature is high enough—some 15,000,-000° C., as was first worked out by Eddington in the early 1920s—for atoms to smash together with tremendous force. Most of the sun is hydrogen, and hydrogen atoms, in smashing together, fuse into the slightly larger atoms of helium. The process is called "hydrogen fusion."

The helium formed in the course of hydrogen fusion has a mass that is 0.71 percent less than the hydrogen out of which it was formed. That mass loss is converted into energy.* Physicists know how much energy is represented by that mass loss and calculate that 588,000,000,000 kilograms of hydrogen must be converted into helium *every second* in order for the sun to keep on radiating energy at its present rate.

This sounds disastrous, as though the hydrogen supply of the sun couldn't last very long. However, 588,000,000,000 kilograms of hydrogen represents only $3 \times 10^{-18}$ (that is, three quintillionths) of the mass of the sun.

If the sun had begun as a ball of pure hydrogen and if it had been converting hydrogen to helium at the present rate for all the 4.6 eons it has so far existed, only one-twentieth of the sun's supply would so

---

* Mass is a form of energy, so that the conversion does not violate the law of conservation of energy, something Einstein first pointed out in his Special Theory of Relativity in 1905.

far have been consumed, and the sun could continue at the same rate for about eighty-seven additional eons.

However, it doesn't quite work out that way. The sun did *not* begin as a ball of pure hydrogen, because, for one thing, when the big bang took place, it formed a universe that contained a considerable quantity of helium to begin with. By the time the sun was formed, ten eons after the big bang, there was still less hydrogen and still more helium.

When the sun was formed, therefore, it seems to have been 80 percent hydrogen and 20 percent helium, and at present it is 75 percent hydrogen and 25 percent helium. If the hydrogen that now remains were to continue to be consumed at the present rate till it was all gone, then the sun would continue to shine as it does for perhaps sixty-eight additional eons.

Unfortunately, that, too, is an overestimate, for what counts is what happens in the core of the sun, where the temperature is hot enough for fusion, and not in the sun generally. As fusion continues, the core of the sun becomes far richer in helium than the sun as a whole is. The core becomes denser and denser, contracts, and becomes hotter and hotter. Eventually, even though the rest of the sun remains much as it is now, the core will become hot enough to force helium to fuse into still more massive forms of matter, and the heat thus formed will force the sun to expand enormously.

The expansion will cool down the surface of the sun to a mere redheat, but the total radiation given off by the much larger surface (for the sun will expand to a million times its present volume) will be enormous and will, of course, destroy Earth. In this expanded state, the sun will be a "red giant."

This change is not just a matter of theory. There are a number of red giants in the sky—enormous stars spread out thinly over a vast volume and gleaming redly. Betelgeuse, Antares, and Mira are the three best-known examples.

After the sun, or any star, expands into a red giant, it goes through other phases, all of which are relatively short compared to the long time it had remained on the "main sequence"—that is, the time during which hydrogen fusion is the chief source of energy, as is true of the sun right now.

What we must ask, therefore, is: How long does a star, and in particular the sun, remain on the main sequence? It is only during that period that Earth can remain a habitable world.

It turns out that the length of time a star remains on the main sequence depends on its mass. The larger the mass of a star, the larger the mass of its hydrogen. However, the larger the mass of a star, the greater the rate at which hydrogen must undergo fusion in order to heat the star to the point where it remains expanded and doesn't collapse under its own gravitational pull.

The rate at which hydrogen must undergo fusion increases at a much faster rate, as a star increases in mass, than the hydrogen supply does. This means that the more massive a star, the quicker its hydrogen consumption reaches the point where the red-giant stage begins. The more massive a star, in other words, the shorter its stay on the main sequence.

The most massive stars known will remain on the main sequence only 1,000,000 years or less. Sirius, which is the brightest star in our sky and which is moderately large, being 2.5 times as massive as the sun, pays for its mass by remaining on the main sequence for a total of only 500,000,000 years—half an eon.

If the sun were as massive as Sirius, it would have entered the red-giant phase over 4.1 eons ago, and life on Earth would scarcely have had a chance to produce even its most primitive cellular form before Earth became uninhabitable. (Naturally, Earth would have had to be more distant from this more-massive sun than it is from the actual sun, or the planet might not have been habitable in the first place.)

The sun, considering its actual mass, will remain on the main sequence, it is calculated, for a total of about twelve eons at most. Since it has already existed for 4.6 eons, that leaves 7.4 eons remaining. It will slowly get hotter during that period, and the last eon or so may be too hot for life on Earth.

We can consider the sun, then, to be a star in its middle age, with the time remaining to it as a life-warming object only a little more than that which has already elapsed. Even if humanity (and its evolved successors) manages to preserve Earth through all its crises, Earth as a habitable world must come to an end after, let us say, 6,000,000,000 years.

## THE DEATH OF THE UNIVERSE

But the sun is only one star out of up to, perhaps, 300,000,000,000 stars in our own galaxy, plus enormous numbers, in addition, in the up to 100,000,000,000 other galaxies that exist. It may be that long before Earth, as a habitable planet, comes to its end, humanity may have spread out to planets circling some of these other stars. Or, failing that, it may be that there are other intelligent life forms, either existing now or to be evolving in the future, that may still be flourishing after we have come to an end.

In fact, we need not be so intelligence-centered as to be concerned with whether intelligent life (or any life) exists or does not exist. The stars themselves will be shining, and perhaps that, in itself, is enough to mark the universe as a whole as still alive. How long will the stars in general shine after our sun has lived out its evanescent life and is gone?

To be sure, by the time that our sun has left the main sequence, all stars that now exist and are more massive than the sun will have left the main sequence earlier—but only about 4 percent of the stars in the galaxy are more massive than the sun. Another 9 percent are roughly as massive as the sun and will leave the main sequence a little before or a little after, depending on when they were first ignited.

That leaves about 87 percent of the stars in our galaxy (and, presumably, in all galaxies) that are distinctly less massive than the sun and that will remain in the main sequence for considerably longer than twelve eons. Most of them have main-sequence lifetimes considerably longer than the period that has elapsed since the big bang and may well have come to birth as stars in the early days of the universe and yet still be in their comparative youth.

The least massive stars that exist, stars so small that they produce just enough pressure and temperature at their cores to ignite the nuclear fires, and that shine only with a dim red light (they are the smaller "red dwarfs"), consume their hydrogen at so parsimonious a rate that despite the comparatively small supply of fusible material, they may remain on the main sequence for as long as 200 eons.

But this takes into account only the stars that exist now. Surely, new stars are continually forming. The sun itself was formed less than five eons ago when the universe was already ten eons old, and there are stars forming now even as we watch. Will not stars continue to form for indefinite periods in the future so that even after 200 eons, when the last star that now exists leaves the main sequence, the galaxies will still be ablaze with other stars, none of which exist now?

It is true that stars have been formed in numbers all through the existence of the universe, are being formed now, and will continue to be formed in the future. The possibilities for this happening may, however, decrease steadily.

The chief source of new star formation is the vast clouds of dust and gas that exist here and there in the universe. The supply, however, vast as it is, is limited. There are "elliptical galaxies" with virtually no dust and gas to speak of. Even in "spiral galaxies" (such as our own), which are rich in dust and gas, this richness is confined to the spiral arms. Neither the core nor the globular clusters, which, all together, make up nine-tenths of the mass of such a galaxy, contain much dust and gas.

The number of stars that may form in the future out of this raw material may therefore be only a few percent of all that have formed in the past.

That would still represent perhaps 10,000,000,000 stars in our own galaxy alone, but though this seems like a large number, it is small compared to those stars which now exist. After 200 eons have passed, and all the stars now existing are off the main sequence, the brightness of the galaxy (which depends almost entirely on main-sequence stars) will be only 1/30 of what it now is, and after 400 eons have passed, only 1/1,000 of what it now is.

There are ways in which stars can turn back, at least in part, to gas and dust, however.

After a star turns to a red giant, fusion reactions continue in its core, as more and more complex elements are built up and as the basic hydrogen supply continues to decrease. Sooner or later, the core becomes heavy with iron. Once iron is formed in the core, there is no chance of any further nuclear reaction delivering energy there. Iron is, so to speak, the ultimate ash of stellar activity.

This means that there isn't enough energy produced in the till then actively fusing core to keep the star expanded against its own gravitational pull, and it contracts rapidly. In fact, it collapses.

As the star collapses, its kinetic energy turns to heat, as Helmholtz had pointed out, and its temperature shoots up rapidly to enormous heights. Most of the hydrogen fuel, which even at this stage is still plentiful outside the core, now undergoes fusion. The outburst of brilliance soon dies, however, and the collapsed star quietly cools off, for it has no other source of heat.

The more massive the star, the more catastrophic the collapse, the greater the momentary brilliance, and the tinier the final state.

Stars with the mass of the sun or less will collapse relatively quietly and lose little mass in the process. They will form a so-called "white dwarf," which will be the size of a small planet, though it will retain all the mass of the star, since the atoms will collapse and their fragments will be pushed much closer together than they would be if the atoms had remained intact.

The white dwarf is "white" because it blazes white-hot at the surface immediately after the collapse, radiating far more heat per square meter than an ordinary star would. The rate of radiation dies off, however, rapidly at first, then more and more slowly. It will take eons for a white dwarf to stop emitting visible light and to grow cool enough to be dark, but that will eventually happen.

Yet however hot and radiant a white dwarf is for each unit area of its surface, it is so small that the total amount of light it gives off is much less than that of an ordinary star. Once the sun collapses into a white dwarf, as it will after its red-giant stage, its total brightness will be only 1/10,000 what it is now. That would be the case near the beginning of the white dwarf's life, and the brightness would steadily shrink further as it grew older.

Stars that are distinctly more massive than the sun collapse so catastrophically that they explode and blow off some of their mass into interstellar space. The amount so lost can be anywhere from a couple of percent for stars not very much larger than the sun to 95 percent for the real giants that are fifty times as massive as the sun. Such explosions result in "supernovas," and these are steadily appearing here and there in the universe, some of them temporarily shining as brightly as a whole galaxy of ordinary stars.

The supernovas, after the explosion, collapse into objects far more condensed than white dwarfs, and far less luminous still. They condense to neutron stars or black holes, which may be considered as not contributing to the brightness of a galaxy at all.

Supernovas contribute to the dust and gas supply of the universe, and are important to star formation in three ways.

1. They add to the raw material, of course.

2. The shock wave of the explosion may initiate a wave of compression in a cloud of dust and gas that may happen to exist near the supernova. That sets in motion the formation of a star, or even the formation of a whole cluster of stars.

3. The dust and gas of the supernova is rich in massive atoms that formed at the core of the star, and these are crucial to the eventual formation of life.

Any atom other than hydrogen or helium, the two simplest, was not formed at the time of the big bang, but was formed in the interior of some star and was spread through space as the result of a stellar explosion. About 1.5 percent of the sun's mass consists of atoms other than hydrogen and helium, and these came from the explosion of stars that existed, and exploded, before the sun and its planetary family were formed. All of Earth, except for a small quantity of hydrogen and a trace of helium, is formed of matter that was once at the cores of stars, and this is true also of all the atoms in living tissue (including our own) that is not hydrogen—which means 90 percent of the total mass.

Nevertheless, supernovas by no means guarantee the indefinite continuation of a shining universe.

First, the number of stars that are massive enough to explode and add to the dust and gas of the universe is comparatively small. Less than 1 percent of the whole are massive enough to make a really spectacular explosion (though that's still 3,000,000,000 stars in our own galaxy alone—and billions others existed in the past, and many will yet exist in the future).

Second, the dust and gas they spread through space as they explode is comparatively poor in hydrogen and rich in the more massive nuclei.

The more massive elements may be crucial to the formation of earthlike planets and essential to the formation of life, but they are *not* important fusion fuel. The only important fusion fuel is hydrogen, and that is constantly being used up and is *not* restored, not even by a supernova explosion.

In fact, the clouds of dust and gas that are condensing into stars as the universe ages are growing constantly poorer in hydrogen and richer in other elements, so the stars that form have less potential time on the main sequence.

Despite new-star formation and old-star explosion, our galaxy, and all galaxies, are gradually dimming, and after, say, 400 eons, they will, from the standpoint of brilliance, be virtually dead. Galaxies will consist of cooling white dwarfs, of neutron stars, of growing black holes, and, of course, of an indefinite number of small, cold planetary bodies.

Of main-sequence stars, shining brilliantly like our sun, there will be the barest scattering. If we imagined someone looking at our universe as it is now—at all the scattering of a hundred billion galaxies with their total of ten billion trillion stars—and then looking at the universe of 400 eons hence, with its occasional lonely spark of light, he could be excused if he thought the universe had died.

The situation would be even worse if the universe, which is now expanding, were to continue to expand indefinitely. Each dying galactic cluster would then continue to be moving farther and farther from every other dying galactic cluster, so that 400 eons hence, what sparks of light remained would be spread some twenty-five times as far apart as they would be if the universe remained as it is today. The darkness would, therefore, be that much deeper.

## THE CONTRACTING UNIVERSE

But *will* the expansion of the universe continue forever?

The universe is expanding against the overall gravitational pull of the matter in the universe, and the rate of expansion is therefore slowing. However, the gravitational pull is also weakening as the universe is expanding outward.

What is happening is analogous to what happens when a rocket moves upward against Earth's gravity. If the rocket begins at a high enough speed, one that is greater than that of Earth's escape velocity, it will manage to stay ahead of gravitation, so to speak. The speed slows, but the weakening gravitational pull will never quite slow it to a halt.

If the universe was hurled outward, by the initial big bang, with a speed greater than the universe's escape velocity, it will expand forever. If the speed was less than the escape velocity, the expansion will eventually come to a halt and the universe will begin to contract again.

If the universe contracts again, the rate of contraction will steadily increase until everything in the universe comes together in a "big crunch." Having presumably begun as an object that was incredibly tiny and incredibly hot, the universe may end the same way. Nor will anything have been lost in the process. Not only will all the matter of the universe come together again, but all the immaterial radiation as well.

It may be that the big crunch will trigger a new big bang that will form a new expanding universe of hydrogen and helium, so that the whole process starts over. In that case, we may be speaking of an infinite number of successive universes, each separated from the one before and the one after by a big crunch/bang.

How far apart in time would the big crunch/bangs be in such a case?

It is hard to tell. It depends on how far below escape velocity the universal expansion is. The farther below escape velocity it is, the shorter the interval between big crunch/bangs.

The interval has to be more than thirty eons, of course, since it is now fifteen eons since the last big crunch/bang and the universe is still expanding.

Let us suppose that the interval is 1,000 eons, or 1,000,000,000,000 years.

In that case, after 400 eons, the universe might seem dead, but that would be only seeming. Its rate of expansion would have been slowed to a crawl and, after 500 eons, it would have come to a momentary halt, though for eons before and after that halt any motion (first outward, then inward) would be too small to detect.

Slowly, through the passing eons after that great stand-still point, the universe would contract faster and faster, until the time came for a big crunch and then another big bang blazing outward.

But will this happen at all? Might it not be that the universe is "open"; that the speed of expansion at the time of the big bang was above the universe's escape velocity, so that the universe will expand forever, and will not only die, but will remain dead?

It is a moot point. Astronomers estimate that in order for the rate of expansion of the universe to be below escape velocity, the average density of the universe (if the matter in all the stars and other objects is smeared out evenly) must be equivalent to about 3 hydrogen atoms per cubic meter. That would account for a gravitational field just intense enough to halt the expansion eventually, and "close" the universe.

If the average density were still greater, the expansion would be stopped more quickly. The higher the average density, the shorter the interval between big crunch/bangs.

In actual fact, though, astronomers making the best estimate they can of the average density of the universe find that that average is only about 1 percent of what is needed to close the universe. If that is so, then there just isn't enough gravitational pull to stop the expansion. The universe is open, will expand forever, will die, and will remain dead.

Suppose, though, that we consider the actual makeup of the universe. The atoms that make up the universe are, in turn, made up chiefly of particles called "protons," "neutrons," and "electrons" (to which we will return later in the book). In addition, there are "photons," which are the fundamental units of light and other similar radiation. Finally, there are tiny particles called "neutrinos."

If all the matter in the universe were smeared out equally and if we took a volume equal to 30 cubic meters, the best estimates now are that that volume will contain 1 proton, 1 electron, 1,000,000,000 photons, and 3,000,000,000 neutrinos.

Let's consider each of these separately. In the place of a proton, there's a 1 in 10 chance that there may be a neutron instead. However, a neutron has just about the mass of a proton, so that this would not affect the average density, and we can just speak of protons, for simplicity's sake.

The electron is only 1/1,836 as massive as the proton, so its presence or absence affects the average density of the universe by only 1/20 of 1 percent, and we can ignore it.

The photon has no "rest-mass"; that is, it would have zero mass if it were at rest relative to the universe generally. It is, however, moving very rapidly and has energy content. This energy content is itself equivalent to a certain amount of mass, but not much. The mass represented by the energy content of a particle is far below the rest-mass of a particle like the proton. In fact, the energy of all the billion photons in the 30 cubic meters is far less in mass equivalence than the rest-mass of the one proton. The photons can therefore be ignored as far as determining the intensity of the gravitational field of the universe is concerned.

The same may be said of the neutrino as of the proton, and it, too, could be ignored.

We could then conclude that of everything in the 30 cubic meters, the only thing we need count is the one proton—and that is not enough. We would need at least a hundred protons in that volume if expansion is to be stopped and the universe is to be considered closed.

In 1980, however, the American physicist Frederick Reines (1918– ) announced the results of certain experiments that made it seem that the neutrino did have a very small quantity of rest-mass after all. Some estimates would make it seem that it has a mass equal to 1/13,000 that of an electron, or about 1/23,000,000 that of a proton.

If this is so (and the experiments that make it seem so are very controversial as yet), then the 3,000,000,000 neutrinos in the 30 cubic meters would have a total mass equal to that of 125 protons. This would be enough, and more than enough, to close the universe and ensure that it will some day contract and that a new, young universe may eventually come into existence.

As we see, then, the horizon of time has been stretched backward fifteen eons into the past, to the birth of the universe; and forward into the future for ten to a hundred times that far, to the death of the universe—or to its rebirth.

# 16
# Instants of Time

## TO THE SECOND AND BEYOND

In probing the horizons of time, both past and future, we have been stretching our minds into the incredibly long. There is also a horizon, however, involving incredibly short stretches of time, and, in some ways, it was much easier, prior to modern times, to deal with years and centuries than with hours and seconds.

In ancient times, the problem of dealing with fractions of a day arose out of the fact that there seemed no natural cyclic changes with a period less than a day. The best that could be done, at first, was to judge which portion of the sky the sun was in, and to speak of dawn, sunrise, morning, midday, afternoon, evening, twilight, and night.

Staring up at the sun to judge its position is neither comfortable nor safe. It is much easier to drive a stake into the ground and study the shadow. In the North Temperate Zone, where the early civilizations arose, the sun rises in the east and crosses the sky (remaining always south of the zenith) and sets in the west. The shadow points west in the morning, and grows shorter as the day progresses. It is shortest at noon, when it points due north, then lengthens eastward as the day wanes toward its close.

Such a "sundial" was used in Egypt as early as 3500 B.C. Over the centuries, the devices grew more complex, and very elaborate sundials were eventually used which made it easy to divide the day into portions of roughly equal duration. It became conventional to divide the day into twelve "hours" (from a Greek word, meaning "time of day").

Twelve was used because it was easy to divide that number evenly

227

by two, three, four, and six. In imitation, the night was also divided into twelve hours (even though sundials are useless for determining the passage of time when the sun is not in the sky), thus making twenty-four hours to the day.

Since, away from the equator, the days grow longer and shorter in a yearly cycle, and the nights as well, in the opposite sense, the equal division of the two periods meant that daytime hours were longer in the summer than in the winter and nighttime hours were longer in the winter than in the summer.

This was inconvenient because the internal time sense didn't change with the seasons. One got hungry and tired at similar intervals, winter or summer. Fortunately, there were other ways of estimating time passage, such as setting up a very slow movement and following its progress.

Sand, pouring through a narrow orifice, might take one hour to complete a passage from an upper to a lower chamber as judged by a sundial at equinox (when the day and night are equal in length). Such an "hourglass" would measure an hour equally, any time of the year. If a sundial is turned over each time it completes an hour, it will count off the hours consecutively and equally, all day long and all night long.

The advantage of an hourglass is that it works at night as well as at day, and when the sky is cloudy as well as when it is clear.

Or else, lamps or candles could be burned and the passage of time judged by the lowering of the level of oil or the burning away of the candle.

The most successful device of this sort was the "water clock" or "clepsydra" (from Greek words meaning "hidden water"). In such clocks, water dripped out of one reservoir into another. From the fall of the water level in the first and the rise in the second, the passage of time could be judged. Ingenious methods were used to make sure the drip was constant with time. Pointers on floats were used to mark off the hours.

The Egyptians started using clepsydras about 1400 B.C. at the latest, and the Greeks borrowed it from them about 150 B.C.

Some time in the 1300s, mechanical clocks came to be used. Pointers were made to turn around dials by means of moving gears. The gears were notched and weights pulling downward forced the gears to turn from notch to notch in regular intervals of time. Since the arrangement was such that the gears stopped, then escaped from confinement and moved only to be stopped again, then did the same over and over, that sort of device was called an "escapement."

Mechanical clocks were steadily improved throughout the Middle

Ages, and for the first time, clocks could be relied on to measure fractions of an hour. (The hour, by a convention first worked out by the Sumerians in the Tigris-Euphrates valley, prior to 2000 B.C., was divided into sixty minutes. Again, sixty is a convenient number because it can be divided evenly in many ways. "Minute" is from a Latin word for "a small part.")

The best mechanical clocks of the medieval variety might keep time to within five minutes a day.

A turning point came in 1581, when an Italian teenager, Galileo Galilei (1564-1642), happened to be watching the swinging chandelier in the Cathedral of Pisa during services. He timed the swings by counting his pulse beat and noticed that it seemed to take the same time for the chandelier to move from side to side whether it swung through a long arc or a short one. He tested this at home by suspending weights and having them swing through arcs of various size.

For the first time, a cyclic motion of constant period was found for small time intervals.

As it turned out, the pendulum does not have an absolutely constant swing. It takes a little more time for long swings than for short ones. About 1657, however, the Dutch astronomer Christiaan Huygens (1629-95) devised a way of having the pendulum swing through the arc of a cycloid, a kind of flattened semicircle, rather than that of a circle. Under this condition, the swing was indeed constant. Huygens then devised a clock in which slowly falling weights kept a pendulum swinging, and the constant swing moved the escapement.

He thus invented what we call today the "grandfather clock," using a long pendulum to beat out seconds. (Each second is 1/60 of a minute. It is called a "second" because it is the second division of the hour, being 1/60th of 1/60th of an hour.)

Huygens's first pendulum clock didn't gain or lose more than ten seconds a day, and so it was thirty times as accurate as the very best mechanical clock. By 1730 such clocks had been improved to the point where they kept time to one second per day; by 1830 to 1/10 of a second per day; by 1885 to 1/100 of a second per day; and by 1925 to 1/5,000 of a second per day.

In 1675, the English physicist Robert Hooke (1635-1703) discovered that fine spiral springs ("hairsprings") could expand and contract about an equilibrium position in constant times. By using a strong, uncoiling "mainspring" in place of weights and a hairspring in place of a pendulum, a portable timepiece could be manufactured, the first useful "pocket watch."

The English instrument maker John Harrison (1693-1776) improved such portable devices to the point where one would gain or lose less than a minute after being on board a pitching vessel for five months at sea.

Other devices, such as electric clocks and clocks run by oscillating crystals or by oscillating atoms, have introduced steadily increasing accuracy. There are "atomic clocks" now that do not lose or gain more than one second in 1,000 years.

One might wonder why one would need such refined accuracy in measuring time. What is a one-second discrepancy in an hour or two, let alone in 1,000 years?

In ordinary life, the need for increased accuracy came with advancing industrialization. The advent of the railroad and the development of timetables made it important to tell time to the minute. The advent of the radio made it desirable for listeners to know the time to very nearly the second.

By the 1800s, pocket watches were the rule among prosperous males in industrial nations. After World War I, wristwatches became the rule among a large fraction of the population of both sexes. At the present time, nearly everyone in a nation like the United States is, in one way or another, conscious of the time of day at all times.

It is to scientists, however, that ultra-precise timing is necessary, and it is to them that extremely short time intervals became important.

## HALF-LIVES

Consider, for instance, radioactive half-lives. Uranium-238 has a half-life of 4,500,000,000 years. This isn't measured directly. You can't wait 4.5 eons to see that half the uranium has broken down. You can, however, count the number of breakdowns in a fixed number of uranium-238 atoms over a fixed period of time and, from that, calculate the half-life. Naturally, the more accurate the measurement of that fixed period of time, the more reliable your final figure for the half-life.

The first half-lives encountered were very long, because the first radioactive atoms discovered were those that still existed despite the fact that they had been breaking down on Earth ever since the formation of the planet 4.6 eons ago. In that time, half the original uranium-238 had broken down; enough is still left to make uranium a comparatively rare metal, but not an excessively rare one.

Yet not all half-lives are long. Uranium-238 breaks down, eventually, to lead-206, but it does so through a chain of intermediate radioactive atoms. In any ore containing uranium-238, all the intermediate radioactive atoms are present also.

All the intermediate radioactive atoms are shorter-lived than uranium-238, so they break down faster than they accumulate. That means that there are very small concentrations present of those intermediates. It can be shown that the shorter the half-life, the smaller the concentration of the intermediate, relative to that of uranium-238.

For instance, in rocks that contain uranium-238 there is also an isotope called radium-226. The radium-226 is present in concentrations only 1/2,840,000 that of uranium-238. The half-life of radium-226 is therefore only 1/2,840,000 that of uranium-238, or about 1,620 years. (Actually, it is easier to determine the half-life of radium-226 and, from that, calculate the half-life of uranium-238.)

There are half-lives shorter than that of radium-226, too, and the short half-lives can be calculated in different ways. Thus it was shown that the shorter the half-life of any atom which broke down by emitting certain radiations called "alpha particles," the more energetic those alpha particles would be. The energy of the alpha particles could be determined by noticing how deeply they penetrate matter, and from that their half-lives can then be calculated.

Among the intermediates of uranium-238 breakdown, polonium-210 has a half-life of 138.4 days, bismuth-214 has a half-life of 19.7 minutes, astatine-218 has a half-life of two seconds, and polonium-214 has a half-life of 0.00016 second.

Naturally, these short-lived atoms would not be existing on Earth today if they were not constantly being formed by uranium-238 breakdown.

In the breakdown of thorium-232, an even more unstable intermediate is formed. It is polonium-212, with a half-life of 0.0000003 second. This is only a third of a millionth of a second, and yet the existence of a substance that is so evanescent that it can't be isolated in visible amounts is indicated by the nature of the alpha particles it emits—and its properties can be determined, too.

The human horizon of brief intervals of time has stretched far below the millionth of a second, too. In the 1930s and thereafter, huge "particle accelerators" were built, enabling physicists to work with particles far smaller than atoms ("subatomic particles") and to endow them with enormous energies. By smashing these subatomic particles into atoms and into each other, they produced new varieties of subatomic particles with exceedingly short half-lives.

These evanescent particles come into existence with enough energy to be traveling at nearly the speed of light. They endure so briefly, however, that even at that speed they make very short tracks in those instruments designed to observe them by the track of water droplets or tiny gas bubbles they leave.

Thus, the theta-meson exists, after formation, for only about $10^{-10}$ seconds (a hundred-trillionth of a second), and there exist "resonance particles" whose lifetimes are worked out to be something like $10^{-23}$ seconds (ten trillionths of a trillionth of a second).

And yet even this is not the limit. In the last few years, physicists have been working out the "Grand Unified Theory" which, it is hoped, will deal with all the different types of forces we know * and group them all under a single set of mathematical relationships. By using the GUT, as it is called, physicists have worked out the possible course of events immediately after the big bang. The GUT *may* give meaningful results (it is conceivable) to events taking place as brief a time as $10^{-43}$ seconds after the big bang. This is 0.0000000000000000000000000000000000000000001 seconds, or a ten-millionth of a trillionth of a trillionth of a trillionth of a second.


## TIME TRAVEL

Is there any way of traveling through time?

In a sense, we do that steadily. Every one of us travels forward through time at the rate of one second per second.

The question is, though: Can we change that rate?

We can if we are in motion relative to the universe generally. The change in rate is small at ordinary speeds. A person traveling in a supersonic jet plane around the world at the rate of 1 kilometer per second for ten years according to his own meticulously accurate clock will emerge at the end of that interval to find that, to the stay-at-homes on Earth, ten years and 9.5 days have passed, so that in the interval he has moved 9.5 days into the future.

Naturally, if he moves faster and faster, the discrepancy will grow larger. If he is in a spaceship moving at a speed of 260,000 kilometers per second (seven-eighths the speed of light), time would move just half as quickly for him as for stay-at-homes. Your clock would register ten years (and he would *experience* ten years) while the clocks on Earth registered twenty years (and everyone on Earth *experienced* those twenty years).

If he moves still faster, the discrepancy will continue to grow. If he

---

* Actually, there are only four of these. They are, in the order of discovery, the gravitational interaction, the electromagnetic interaction, and two different nuclear interactions, the strong and the weak.

moves at 295,000 kilometers per second (98.3 percent the speed of light), when ten years passed for him, 54.4 years will have passed on Earth, and so on.

In this way (in theory) he could travel as far into the future as he wished, provided he possessed the energy that would accelerate him to such a speed, and that would allow him to turn and then to decelerate in order to bring him back to Earth. (And provided traveling at such speeds is practical.)

As for moving backward in time, that is quite another thing.

We can *see* back in time; indeed, we can't help but do that. Light always takes time to reach us, however short the distance, and we see an object only as it was when the light left it. If we are standing 3.3 meters away from something, we see that something as it was $10^{-8}$ second (a hundred-millionth of a second) before. (Naturally, that time difference is totally insignificant on the ordinary scale and we consider ourselves to be seeing something "as it is *now.")*

On the other hand, it takes eight minutes for light to reach us from the sun, and so we see the sun as it was eight minutes ago (and if it had disappeared three minutes ago, we wouldn't know that for another five minutes). We see the star Alpha Centauri as it was 4.3 years ago, and the star Arcturus as it was forty years ago.

When we look at the Andromeda galaxy, we see it as it was 2,300,-000 years ago, and when we look at the farthest quasar we can see we see it as it was 10,000,000,000 years ago or so. The light that we see emerging from that distant quasar began its long journey at a time eons before the solar system existed.

Of course, the farther into the past we peer, the less we see, since we must look through greater and greater distances.

If we could travel away from Earth at a speed faster than light, and if we observed Earth as we did so, we would be outracing the light beams that had left Earth, and, in theory, we could watch Earth's history unfolding backward. But then, the farther we traveled and the farther into the past we tried to look, the less we could see, for we would buy that farther and farther backward look at the cost of greater and greater distances.

Could we do as is done in science fiction stories—sit in some device, twiddle some knobs, and step out onto Earth as it will be in A.D. 3000 or as it was in 3000 B.C.?

This seems exceedingly unlikely, ever. Earth is moving about the sun at a speed of 30 kilometers per second. In three hours, it travels 324,000 kilometers. If we were to step into a time machine, push a button, and step out three hours into the past, then we would find ourselves in outer space, with Earth 324,000 kilometers away.

If we imagine that the time machine would move with Earth and that we could push a button and step out three hours into the past and still find ourselves on Earth, then we would have had to move the 324,000 kilometers in just the time it took to push the button and we would then have been going faster than the speed of light.

Then, of course, it is more complicated than that too, for not only is Earth moving about the sun, but it is moving with the sun about the center of the galaxy and with the galaxy in a path within the local group of galaxies and among the galaxies generally.

If we managed to match all those movements, somehow, then it seems logical to suppose that it would take energy to do so, so that traveling through time would require as much energy as traveling through space. And as much time, too—which would instantly negate the whole idea.

Furthermore, there are paradoxes involved. The old science fictional chestnut of the man who goes back into the past and kills his own grandfather, so that he himself was never born and could not have done the deed, is merely the most-cited case of the way in which time travel into the past would destroy the principle of causality.

Finally, even if the principle of causality could be violated, and if we could imagine that some technological advance in the future would make time travel practical, it still remains that, as far as we know, no one from the future has come to visit us.

That might mean that the time travelers are careful not to let themselves be seen—or it might mean that no one in the future will ever solve the problem. (I suspect probably the latter.)

# 17
# Speed

## LIVING THINGS

Before moving on further, let us combine the two areas we have already covered—those of space and time. It is possible to move through space in a given time, as we have had occasion to take for granted throughout this book.

The rate of displacement—that is, the distance one travels through space in a given time—is the "speed." (If one considers speed in a given direction, one can speak of "velocity," a more precise term. For purposes of this chapter, however, we need speak only of speed. The direction of displacement is immaterial.)

Human beings have always been able to move in one of two chief ways: walking and running. In walking, at least one foot remains on the ground at all times, while in running, there are times when both feet are off the ground. (Jogging is a kind of slow run, intended for endurance or exercise. It is also possible to progress by hopping, skipping, turning cartwheels, and so on, but those are games or exhibitions and are not normal modes of progression.)

A normal brisk walk might be at the rate of 5 kilometers per hour, though this speed can be raised when necessary. Generally, when greater speed is needed human beings break into a run, but fast walking can be maintained over sizable distances, usually as exercise, or in sporting competitions.

To those who do not have occasion to undertake walks that are either long or rapid, it is surprising to discover what the human machine can accomplish through training and practice. The world record for walking 20 kilometers is 1 hour, 25 minutes, and 19.4 seconds, which amounts to a speed of 14 kilometers per hour.

Naturally, as distance increases, the average speed drops. The record for 50 kilometers is 4 hours, 27.2 seconds, a speed of 12.5 kilometers per hour.

Running is, all things being equal, faster than walking. The most famous long race is the "marathon," so called because it is supposed to represent the distance covered by the Athenian runner, Pheidippides, in reporting the result of the Battle of Marathon to the Athenian people in 490 B.C. The length of the marathon is 42.2 kilometers, and the record time is 2 hours, 8 minutes, 33.6 seconds. This represents an average speed of 19.7 kilometers per hour, which is roughly 1.6 times as fast as the speed with which the same distance could be negotiated by an expert walker.

Naturally, over shorter distances, the average speed goes up. The record for the 10-kilometer run is 27 minutes, 30.8 seconds, for an average speed of 21.8 kilometers per hour. The 1-mile run (1.6 kilometers) has been negotiated in 3 minutes 51.1 seconds, for an average speed of 25 kilometers per hour. The record for the 100-yard dash (0.0914 kilometers) is 9.0 seconds, for an average speed of 36.56 kilometers per hour.

Of course, the average speed of an athlete running the 100-yard dash is reduced by having to accelerate from rest to begin with. In 1963, a runner was timed between the 60-yard and 75-yard marks and over that 15-yard distance he achieved a speed of 44.9 kilometers per hour, perhaps the fastest speed at which a human body has hurtled, unaided, through space for a significant distance.

Through most of human existence, human beings have had to depend on their own muscles for progression over distances. Horses were tamed by Asian nomads abut 2000 B.C., and thereafter these animals were increasingly used by human beings who wished to travel at speeds greater than they themselves could maintain (or who wished to travel with less personal effort expended).

Under ordinary use, ordinary horses can move more quickly than ordinary men can, and horses, especially ones bred for speed can, under favorable conditions, far outdo human racers over similar distances.

The world record for a 3-mile run (4.83 kilometers) by a racehorse is 5 minutes 15 seconds, which represents an average speed of 55.2 kilometers per hour. Over shorter distances, horses can attain speeds of 64 kilometers per hour, which is nearly half again as fast as the fastest speed at which a human being can run, and this can be maintained over distances of 1.5 kilometers.

There are not many land animals that can outdo a racehorse. A jackrabbit, running at maximum, can perhaps keep up with a fast racehorse, thanks to its ability to leap long distances. A large kangaroo, frightened and running for its life, can make even longer leaps (one of 12.8 meters has been recorded) and can overtake a racehorse.

Oddly enough, one two-legged animal can also do this. An ostrich, when running for its life, is thought to be capable of reaching a speed of 70 kilometers per hour. There is no question that it is the fastest living organism that runs along the ground on two legs.

Many species of deer, antelopes, and gazelles can outrace a horse. The red deer has been reported to reach speeds of 67 kilometers per hour. The pronghorn antelope can apparently reach speeds of nearly 100 kilometers per hour when it fully extends itself.

These land animals I have mentioned are all herbivores. They live on flatlands, and their one sure escape from carnivorous predators is speed. The swiftest are most likely to survive, and the course of evolution has produced animals that can race across ground amazingly well.

Carnivores, by and large, do not run as quickly as their prey. They can manage to find their dinner by lying in wait, or by stalking, or by springing from ambush. They can also make do with the young, the old, the ill, the hurt—animals that for one reason or another cannot keep up with the herd at top speed.

Nevertheless, some carnivores *do* speed along. The coyote can manage 55 kilometers per hour if pressed, perhaps not *quite* as fast as a jackrabbit, but fast enough.

And yet the fastest of all land animals is a carnivore, after all—the cheetah. Over a distance of half a kilometer, it can maintain a speed of up to as much as 102 kilometers per hour. It runs down even gazelles, but only if it can do so in a very short time indeed (or if the gazelle is, for some reason, hampered in its running), for if the cheetah hasn't seized its prey within a minute, it has to slacken off, and the gazelle, though not quite as fast as the cheetah, has greater endurance and can keep running.

Human beings can progress through water, but by no means as quickly as on land. Water is more viscous than air and it takes more energy, all else being equal, to force one's way through water than through air. Anyone who races across the beach and into the water at a full run finds this out at once.

The greatest speed recorded for a human in the water is only about 8.1 kilometers per hour.

Human beings, however, are far from being well adapted to

movement in the sea. Other warm-blooded animals that have returned to the sea, and live there for much or all of the time, are better adapted. They are more or less streamlined to minimize water's resistance, and they have flippers, paddles, or flukes to serve as propulsive devices.

Nevertheless, the fastest dolphins, seals, and penguins do not exceed speeds of 40 kilometers per hour.

By far the most successful and best-adapted of the sea animals are the fish. It is not surprising, therefore, that the fastest water travelers are to be found among that group, particularly among the large and muscular ones.

The great blue shark has been reported to travel at a speed of 65 kilometers per hour, which would make it as fast as the fastest racehorse, and yet it is outdone by others. The marlin is supposed to travel at speeds of up to 80 kilometers per hour, which would make it as fast as a speedy antelope. Sailfish hold the record, though; a speed of 110 kilometers per hour was recorded for one, making it faster than a cheetah.

But if water is more difficult to push through than air, traveling through air directly, without touching the ground, offers greater chances at speed than progression on the ground does. All things being equal, in other words, it is faster to fly than to run.

That is part of the explanation of the speed of animals with long strides, as in the case of horses or ostriches, or with the ability to make long hops, as in the case of jackrabbits and kangaroos. These are quasi-flyers.

If we talk of true flyers, however, we find that the list includes four different types of animals.

Of these, the pterosaurs have been extinct for 65,000,000 years, and we do not know how fast they might have been able to fly. There seems to be a general impression among paleontologists that they were not fast flyers.

As for insects, it is very difficult to tell how fast they truly fly, and how much of their speed is contributed by air currents. It seems extremely doubtful that under ordinary circumstances any insect can fly faster than 55 kilometers per hour and, even then, only briefly, or at heights where the air is thin and offers less resistance.*

Dragonflies are among the fastest insects, and an extinct dragonfly that lived about 300,000,000 years ago appears to have had the enor-

---

* There have, in the past, been reports that certain insects could fly at speeds of up to 1,000 kilometers per hour, but such claims have been shown to be clearly impossible. In reporting extremes of any sort, particularly about living organisms, there is always a tendency to exaggerate for dramatic effect. I have tried to be conservative in this book, but there's a chance I have fallen victim here or there.

mous wingspread (for an insect) of 0.7 meters. It may possibly have been capable of bursts of speed of up to 70 kilometers per hour.

Bats, the only flying mammals, have evolved, apparently, in the direction of maneuverability rather than speed. The highest speed recorded for any bat is 51 kilometers per hour, which means that a fast bat is no faster than a fast insect.

For flight speeds greater than that of a fleet racehorse, we must turn to the birds, which are the speediest of all the flying creatures that have ever evolved.

The swan, for instance, has been clocked at speeds of up to nearly 90 kilometers per hour, and the Canada goose up to nearly 100 kilometers per hour. The Canada goose, in other words, is nearly as swift as a cheetah, and can undoubtedly keep up its speed for much longer.

Yet there are birds that far outdo the Canada goose. The spine-tailed swift has been reported to fly at speeds of up to 170 kilometers per hour, or better than 2.8 kilometers per minute! Assuming this to be so, this swift is the fastest known living organism that depends for speed on its muscles alone.


## INANIMATE OBJECTS

So far, we have discussed the unaided speed of living organisms. The human species, however, does not rely on its muscles alone.

Where travel over water is concerned, for instance, human beings have made use of ships of one sort or another from the early days of civilization. Early ships drifted downstream, making, at best, passive use of inanimate energy. Eventually, human beings learned to pole or row ships in order to make way against the current, and this still made use of human muscle for propulsion.

Rowing was never a fast method of getting around. Even in very light boats, propelled by a team of rowers in racing competition, the highest speed recorded is but 21.6 kilometers per hour.

By outfitting a ship with sails, the wind can be used for propulsion—again passively. As the centuries passed, sails were made more numerous and efficient, and their use more ingenious. Other aspects of ship design were also improved. In 1890, a sailing ship averaged 40 kilometers per hour during the course of a half-day run.

Eventually, the use of inanimate energy produced and maintained by human beings and bent to their will came into use in connection with transportation and speed. This could not happen, of course, until after the introduction of efficient steam engines in 1774 by the

Scottish engineer James Watt (1736-1819), an event that initiated what is called "the Industrial Revolution."

It was not long before a steam engine was placed on board ship and set to the task of turning a paddlewheel for propulsion. In 1790, the American inventor John Fitch (1743-98) built the first successful steamship, but was unable to make a commercial success of it. In 1807, however, Robert Fulton (mentioned earlier in connection with submarines; built the *Clermont,* the first commercially successful steamship.

Steamships were lumbering objects to begin with, and during the first decade of their use they moved at only about 12 kilometers per hour or so. Their advantage was primarily their ability to move in a dead calm, or even against the wind.

Sailing ships held their own through the first six decades of the 1800s, but the coming of ironclads, of steadily more efficient engines, and of screw propellers in place of paddlewheels put steam in a steadily mounting lead from the 1870s on. With the twentieth century, sails were increasingly confined to pleasure craft.

Steamships eventually far outstripped sailing ships in speed. The liner *United States* in 1952 crossed the Atlantic from New York to Le Havre in less than three and a half days, for an average speed of 66 kilometers per hour. Later in the year, she attained an average speed of 77 kilometers per hour over the course of a day's run.

Warships can do better still. The fastest submarines can go 83 kilometers per hour, the fastest destroyer 113 kilometers per hour, and specially designed experimental vessels up to 130 kilometers per hour. Not surprisingly, then, human beings (aided by technology) are now the fastest of all living things at sea.

On land, the use of inanimate energy for transportation lagged far behind the situation on the water. Until well into the 1800s, human beings were confined to muscles, either their own or those of horses, for crossing the kilometers. (The use of other animals, such as camels and donkeys, played a minor role.)

In 1801, the English inventor Richard Trevithick (1771-1833) was the first to place a steam engine on a vehicle and arrange to have the engine's power turn the wheels. Naturally, this wouldn't work on uneven ground, since too much power would be wasted in overcoming friction and irregularities. Trevithick showed that it was practical to have the wheels turn on long metal rails. He thus produced not only the locomotive, but the railroad.

Unfortunately, Trevithick, like Fitch before him, could not make a commercial success of his venture. That was left for the English inventor George Stephenson (1781-1848) in 1825. Thereafter, railroads gained popularity rapidly and between 1850 and 1950 were the dominant means of long-distance land travel.

The first locomotives, in the 1820s, reached speeds of not more than 47 kilometers per hour, but the speed rapidly increased as improvements were added. By 1840, speeds of 91 kilometers per hour were recorded, and, for the first time in history, human beings (aided by technology) could travel overland at speeds considerably greater than that of a racehorse, and could, given enough fuel, keep it up indefinitely. (To be sure, ordinary railroad passengers didn't travel at nearly the speed set by specially designed devices traveling under specially favorable circumstances—but then, neither did ordinary horsemen travel at the speeds of jockeys mounted on thoroughbred racehorses running under specially favorable conditions.)

In 1843, less than twenty years after Stephenson's first practical locomotive, a locomotive in Ireland attained a speed of 137 kilometers per hour and human beings could now outrace cheetahs.

By 1903, electric locomotives in Germany were clocking speeds of 210 kilometers per hour and human beings could travel faster than any other living creature, even the fleetest swift. At the present time, locomotives whose power is augmented by rockets can achieve speeds of something like 380 kilometers per hour.

In 1860, the Belgian-French inventor Jean J. E. Lenoir (1822-1900) designed the first useful internal-combustion engine, which proved, eventually, to be much lighter and more efficient than the steam engine. The internal-combustion engine was particularly adaptable to small vehicles, and in 1860, Lenoir attached one to a wagon and produced the first "horseless wagon," something that was eventually to be called an "automobile."

Automobiles didn't become practical till the 1880s. One of the first true automobiles was built by the German inventor Gottlieb W. Daimler (1834-1900) in 1887.

As the 1900s progressed, the automobile became the dominant form of land transportation for family groups, while larger versions, buses and trucks, became increasingly important in the transportation of larger human groups, and of merchandise of all kinds.

The speed limit on American highways is now 88.5 kilometers per hour (55 miles per hour), and although this is the speed of a gazelle, drivers have the greatest difficulty keeping down to this level. (Once, while traveling from Boston to New York in 1973 under the stress of a family emergency, I averaged 110 kilometers per hour and hit peaks of 130 kilometers per hour. Not only did I far outrace a cheetah, but I maintained it for three hours.)

Auto racers have attained speeds as high as 320 kilometers per hour over the course of a lap. and in 1970. a rocket-assisted ground vehicle achieved a speed of 1.046 kilometers per hour. which is the equivalent of 17.4 kilometers per minute. This approaches the speed of sound. which travels through air. under standard conditions. at a speed of 1.192 kilometers per hour or 19.9 kilometers per minute.

Naturally. airplanes can outpace anything on the ground. even as birds can outpace any land animals. The speed of sound was first exceeded by human beings (or any living thing. for that matter) on October 14. 1947. when an airplane piloted by Charles E. Yeager managed to do so.

Since then. "supersonic flight" has become commonplace. and rocket-assisted airplanes have achieved speeds in excess of 7,300 kilometers per hour. which is the equivalent of 2 kilometers per second and is 6.1 times the speed of sound.

Rockets can. in turn. outpace airplanes. Probes headed for the outer planets leave Earth at speeds of about 14.5 kilometers per second. The space probe Helios-B. moved around the sun in 1976. at a distance of 43.500.000 kilometers from the sun. It attained a speed of 68.4 kilometers per second. These are the fastest man-made objects so far.

With the rocket. we approach the speeds of astronomical objects. and exceed some.

Earth. for instance. turns on its axis in a day. Every point on its surface. therefore. sweeps out a circle about the axis, and the points that sweep out larger circles move at faster speeds relative to the axis. The largest circle is that of the equator, and a point on Earth's equator is moving, relative to Earth's axis, at a speed of 1,670 kilometers per hour, or 0.46 kilometers per second. This is less than 1/30 of the speed of our fastest rockets.

Earth does not hold the rotational speed record for the solar system. however. The sun, which has a circumference 109 times the length of Earth's, rotates in 25.1 days. A point on the sun's equator moves. relative to its axis, at a speed of 7,260 kilometers per hour, or 2 kilometers per second.

That is not the record, either. The planet Jupiter, with a circumference 11.16 times that of Earth, rotates in not much more than two-fifths of the time it takes Earth to do so. A point on Jupiter's equator (or at least on the cloud layer that forms its visible surface) would move at a speed of 45,000 kilometers per hour relative to Jupiter's center, or 12.6 kilometers per second—and even this is slower than our fastest rockets.

In addition to rotating, planets fly through space in their voyage around the sun. These speeds of revolution are greater the closer a planet is to the sun and, therefore, the more intense the sun's gravitational field is at the site of the planet.

Jupiter, for instance, moves around the sun at an average speed (relative to the sun) of 13.1 kilometers per second, a speed that falls just short of our fastest rockets.

Mars, which is considerably closer to the sun than Jupiter is, revolves at an average speed of 24.1 kilometers per second, and that outdoes our rocketry.

Earth's average speed of revolution about the sun is 29.8 kilometers per second, while Mercury, which is the closest planet to the sun and therefore the fastest, travels at an average speed of 47.9 kilometers per second. Mercury has an eccentric orbit; it is considerably closer to the sun at one end of its orbit than at the other. At its closest approach to the sun ("perihelion"), Mercury is moving at a speed of 56 kilometers per second.

There are some comets that come far closer to the sun than Mercury does, and those that approach closest must travel at speeds of something like 600 kilometers per second as they skim by the sun's surface.

Just as planets revolve about the sun, the sun revolves about the center of the galaxy, doing so, relative to that center, at a speed of something like 300 kilometers per second. Stars that are closer to the center than the sun is will, naturally, revolve at greater speeds. (For a star that approaches *too* close to the center, more and more stars will lie farther from the center than it is and the gravitational pull of those stars that are still between itself and the center will decrease, and so the speeds of revolution will decrease again.)

The universe is, of course, expanding, and the galaxies are receding from each other at rates proportional to their separation. During the 1920s and thereafter, as dimmer and dimmer galaxies were studied spectrographically, speeds of recession were determined that were in the range of thousands of kilometers per second, then tens of thousands, then hundreds of thousands. By the early 1960s, distant galaxies were studied which were receding from us at a speed of 240,000 kilometers per second.

Now, at last, we are approaching the speed of light.

The fastest living thing moves at less than 1/6,400,000 of the speed of light. The fastest man-made device moves at less than 1/20,000 of the speed of light. A comet hastening past the sun moves at only 1/500 of the speed of light.

A galaxy receding from us at 285,000 kilometers per second is, however, doing so at a speed that is 80 percent that of light.

In 1963, the "quasars" were discovered. Even the nearest of these turned out to be in excess of 1,000,000,000 light-years distant, farther than the farthest ordinary galaxy that had been detected. Naturally, the quasars were receding from us at speeds even greater than that of ordinary galaxies. The farthest galaxy we have yet detected seems to be receding from us at a speed of 285,000 kilometers per second, or 95 percent that of light.

The only objects that are faster still are some speeding subatomic particles in extremely energetic cosmic rays that travel at better than 99.9 percent of the speed of light, but don't quite reach the critical speed itself.

The speed of light in a vacuum, then, is the absolute speed limit of the universe—299,792.5 kilometers per second.*

---

* In the last few years, astronomers have detected a number of distant quasars in which portions seem to be separating at speeds greater than that of light, but there seems to be complete confidence that there will be explanations forthcoming that will account for such apparent speeds without breaking the speed-of-light limits. As for the hypothetical faster-than-light tachyons, these have not yet been detected, and they may not exist.

# PART III
## The
## Horizons
## of
## Matter

# 18
# Mass, Large and Small

## LIVING GIANTS

So far in this book, we have traveled through space and time, moving out to the limits we can observe or imagine. In the popular mind, though, space and time are only the frame. Within it are objects made of matter. We should search the horizons of matter, as well.

This is more complicated, in a way, than dealing with space and time. Space has one all-important property—extension. Time has one all-important property—duration. In search for extremes in either of these we make use of a single measuring device, the meter-stick for space and the clock for time (though modern technology makes use of incredibly refined versions of each).

In the case of matter there are several different properties, each of which deserves attention. The most important of these is "mass."

Mass is a notion that was conceived by Isaac Newton. He pointed out in the 1680s that different objects require forces of different sizes for acceleration to the same final speed in a given time. The force required is proportional to the mass of the object. It is much easier to accelerate a baseball to a given speed in a given time than an iron ball of the same size (try it and see). We say, therefore, that the iron ball is "more massive" than the baseball.

Again, every piece of matter is the source of a gravitational field. Such fields vary in intensity, some pulling more strongly at objects at a given distance than others do. The intensity of the gravitational field varies according to the mass of the object that is the source of the field.

Einstein, in his theory of relativity, pointed out that mass increases with velocity and reaches infinite value at the speed of light. That is one reason why objects with mass cannot quite reach the speed of light, let alone surpass it.

In this section, though, we will deal only with "rest-mass," the mass an object has when it is at rest relative to the observer who is measuring the mass.

Where rest-mass is concerned, it is reasonable to say that the mass of an object is equivalent to the amount of matter it contains. That is, two identical objects, at rest relative to each other, have twice the mass in total that either has alone.

We measure mass, commonly, by pitting one object against another in opposite pans of a balance. Earth's attraction on the two objects is proportional to the mass of each. When the attraction is equal on each so that the two pans balance and neither moves downward, we know that the hitherto unknown mass of the object in one pan is equal to the known mass of the sum of the "standard weights" we have placed in the other.

Non-scientists commonly speak of the "weight" of an object rather than its "mass" and say that one object is "heavier" or "lighter" rather than "more massive" and "less massive." Strictly speaking, this is improper. Weight and mass are interchangeable at sea level on Earth's surface, but weight grows less with increasing distance from Earth's center and mass does not. In this section, therefore, I will try to remember to speak of mass only, and not of weight.

To scientists, a useful unit of mass is the "kilogram," which I have been using now and then throughout this book. To repeat, a kilogram is equal to 2.2 pounds in common units, or 35.2 ounces. A thousand kilograms is equal to a "tonne," which is equal to 2,205 pounds and, therefore, to 1.1 common tons. In this section I will use only kilograms and tonnes.

Thus, the average mass of an adult male in the United States is about 73 kilograms, and that of an adult female is about 60 kilograms.

Human beings tend to think of objects as large or small in reference to themselves, and large is naturally more impressive. Large animals dispose of more strength and are more dangerous than small, by and large.* Virtually throughout human history, human beings have therefore lived in terror of the large predators.

Even though there is no longer need for such fear in most parts of

---

* This is only a general statement and has obvious exceptions. A rattlesnake is much more dangerous than a cow, and everyone knows it.

the civilized world, and though we encounter the predators only in zoos and circuses, we still associate size and terror, and motion pictures featuring large animals as menaces are almost invariably popular. Naturally, in order to achieve an effect, the menaces must be made large out of all reason. We have only to think of that classic motion picture *King Kong,* with its impossibly magnified gorilla. Almost any impossible magnification will do, whether lobsters, spiders, or dinosaurs.

Legends also speak of human giants. These are found in many folk tales and even in the Bible. Such have never existed, however.

To be sure, there are human beings who have been considerably taller and more massive than average, but the extremes have generally represented hormonal imbalance, and their victims have been short-lived. Whereas the average height of the American male is 1.72 meters, a young man of twenty-two died in 1940 whose height at death was no less than 2.72 meters. Another man, who died at thirty-two, achieved a mass of about 480 kilograms in the course of his life, 6 1/2 times the average. To be sure, the excess mass was largely fat.

If we speak of human beings in the normal range of height and mass, then we can say that modern man is himself the giant of his type. The hominids who were the ancestors of *Homo sapiens* were all smaller than we are.

There is only one living primate that is distinctly larger than the human being, and that is the gorilla, and even the gorilla is no taller than the human being (despite King Kong). However, the male lowland gorilla averages 165 to 180 kilograms in mass, which is up to 2 1/2 times the mass of a male human being.

The largest primate that ever lived belonged to the genus *Gigantopithecus* ("giant ape"), which has been extinct now for at least 2,000,000 years. These were gorillalike creatures who may have stood 2 3/4 meters tall when standing upright, and may have been as massive as 275 kilograms, half again as massive as modern gorillas and nearly four times as massive as men (and still no King Kong).

Human beings have always been familiar, of course, with animals larger and more massive than even the largest primate—horses, bulls, camels. The Bible refers to the "behemoth," which is thought by some to be modeled on the hippopotamus, which is larger than any of these. The largest of all living land animals, however, is the African elephant.

The average adult male African elephant is 3.2 meters tall at the shoulder and has a mass of 5,100 kilograms (5.1 tonnes). The largest African elephant ever measured stood 3.8 meters tall and had a mass of 11 tonnes. Such an elephant has the mass of 150 human beings. (There are extinct elephants that are taller still. Heights of up to 4.8 meters, 2 1/2 times that of a man, have been reported.)

Although no other living land animal can compare in mass to the African elephant, one (and only one) stands taller, and that is the giraffe. The tallest giraffe ever measured had a height of about 5.8 meters (3 1/3 times as tall as the average human male), although it probably weighed only a fifth of what a very large elephant would.

Modern elephants have been outdone in past ages. The largest mammal that ever lived was a rhinoceroslike animal called *Baluchitherium* ("beast from Baluchistan"), which has been extinct for 20,000,000 years. It was taller than a giraffe and more massive than an elephant. It carried the top of its head 8.2 meters above the ground, nearly half again as high as a giraffe's, and it had a mass of some 20 tonnes, which would make it twice as massive as an elephant.

More massive still were some of the extinct giant reptiles which ceased to exist about 65,000,000 years ago. The most massive of these was *Brachiosaurus* ("arm-lizard" because of its long forefeet), and it was, in fact, the most massive land animal of any kind that ever existed. It stood up to 6.4 meters at the shoulder, and the head it carried on its long neck could tower 11.9 meters in the air, which made it twice as tall as the tallest giraffe. Its overall length was about 23 meters, and its mass was about 40 tonnes, which made it twice as massive as a *Baluchitherium* and nearly four times as massive as the largest living elephant.

Sea animals have an advantage over land animals. The buoyancy of water reduces the pull of gravity upon them and doesn't require them to hold up their full mass against that pull or to lift it with each step. Consequently, sea animals can be larger than any land animal.

The largest animals now living are the various species of whales, and the largest of all, living or extinct, is the blue whale. Blue whale females are sometimes more than 30 meters long. The largest which has been measured accurately had a length of 33.3 meters and must have had a mass of something over 135 tonnes. Such a blue whale is almost half again as long as a *Baluchitherium* and 3 1/3 times as massive. For that matter, it is as massive as 1,900 men.

Yet even the blue whale is not the largest living thing, for any animal is outmatched by trees, which are the tallest and most massive living things.

The tallest trees are the redwoods growing in northern California, and the tallest of these which was accurately measured was 112 meters tall. Such a tree is 3 1/3 times as tall as a blue whale is long, and that is so even without counting the tree's root system beneath the ground.

The most massive living things are the sequoia trees. These are shorter than the redwoods but wider. The "General Sherman" sequoia, growing in Sequoia National Park, is only 83 meters tall, but it is estimated to weigh 2,000 tonnes, or nearly 15 times the mass of the largest whale, and as much as 28,000 people. (To be sure, only a small portion of such trees is actually alive. Most of it is dead wood.)

## FROM EARTH TO UNIVERSE

Naturally, there are inanimate objects on Earth that are more massive than any living thing. Some are man-made—the Pyramids of Egypt, for instance. Some are not—Mt. Everest, for instance. The most massive object on Earth is, of course, Earth itself.

Once Isaac Newton had worked out the theory of gravitation in 1687, a method for calculating the mass of Earth existed, at least in theory.

Earth exerted a pull on an object. If one could determine the mass of the object, its distance from the center of Earth, the force exerted upon it, and the gravitational constant, then the mass of Earth could be calculated. Of these four values, three were known or could be easily measured. Only the fourth, the gravitational constant, was unknown. If the gravitational constant could somehow be calculated, the mass of Earth could be at once deduced.

Suppose, then, that the gravitational force between two objects of known mass could be measured. From that force, given the masses of both objects and the distance between them, the gravitational constant could be at once calculated, and the mass of Earth would be known. The difficulty here is that the gravitational force between any two objects small enough to have known masses and to be capable of laboratory manipulation is exceedingly small.

Nevertheless, in 1798, the English scientist Henry Cavendish (1731–1810) managed to make such a measurement and to do it quite accurately, too. He obtained a value for the gravitational constant and, in that way, determined the mass of Earth.

Only minor improvements have had to be made in Cavendish's figure since, and the mass of Earth, as now accepted, is $5.976 \times 10^{21}$ tonnes; that is, 5,976,000,000,000,000,000,000 tonnes. The earth is about as massive, in other words, as three billion billion sequoias of the largest size.

Yet Earth, though eighty-one times as massive as the moon, and ten times as massive as the planet Mars, is small compared to many other astronomical bodies. Even at the time of Cavendish's determination, it was known that there were three planets (Jupiter, Saturn, and Uranus) that were far larger than Earth, and a fourth (Neptune) was discovered in 1846.

All four of these large planets had satellites. If the distance of a particular satellite from its planet can be determined, along with its period of revolution about that planet, and if this is compared with the distance of the moon from Earth and the moon's period of revolution about Earth, then it is possible to calculate the mass of a planet relative to that of Earth.

In this way, it was found that Jupiter was the most massive of the planets by far. It is 317.9 times as massive as Earth and therefore has a mass of $1.9 \times 10^{24}$ tonnes. It is more than twice as massive as everything else in the solar system (excluding the sun) put together.

But we must not exclude the sun, for it is far more massive than even Jupiter. From the distance of Earth from the sun, and from the period of Earth's revolution, we can calculate the sun's mass relative to that of Jupiter.

The sun, it turns out, is 1,019 times as massive as Jupiter and, therefore, 324,000 times as massive as Earth. The sun's mass is $1.99 \times 10^{27}$ tonnes, and it contains 99.9 percent of the mass of the solar system. Jupiter, Earth, all the other planets, satellites, asteroids, comets, and miscellany make up the remaining 0.1 percent.

What about the stars? Are some of them perhaps more massive than the sun?

In 1793, the German-English astronomer William Herschel (1738–1822) discovered that there exist such things as "binary stars"—systems of two stars circling each other. If their distances from us were known, the actual distance of separation of the two stars could be calculated, and from that, and from their period of revolution, their total mass could be determined.

It was not until the 1830s that the distance of any star was determined and the possibility of determining the mass of at least some stars grew better.

In 1844, for instance, Bessel discovered that Sirius had a companion star that was too faint for him to make out. He knew it was there, however, because Sirius was moving in a tiny orbit about *something*. In 1862, the American instrument maker Alvan Graham Clark finally spotted the dim companion.

From the distance of Sirius, from the separation between Sirius and its dim companion, and from the comparative size of the orbits of the two stars, it was possible to calculate the mass of both stars. Sirius's dim companion is about 1.05 times the mass of the sun, while Sirius itself is 2.14 times the mass of the sun.

The sun, therefore, is not the most massive object in the universe. Nevertheless, stars more massive than the sun are considerably less common than those less massive.

In 1924, the English astronomer Arthur Stanley Eddington (1881–1944) demonstrated a relationship between the mass and the luminosity of a star. The more massive a star, the more luminous, with the luminosity shooting upward much more rapidly than the mass. If a star were very massive, it would be producing energy at such a rate that it would tear itself apart.

About as massive as a star can get and stay in one piece, except under very rare and unusual conditions, is about seventy times the sun's mass. A particularly massive star was discovered in 1920 by the Canadian astronomer John Stanley Plaskett (1865–1941). It is at least fifty-five times as massive as the sun, or $1.1 \times 10^{29}$ tonnes.

William Herschel was the first to recognize that the stars visible in the skies form a flat, lens-shaped system, but it took over a century to work out the actual dimensions of this "Milky Way galaxy." From the distance of the sun from the center of the galaxy and the period of revolution of the sun about that center, we can determine the overall mass of the galaxy. The galaxy is now thought to have a mass equal to 140,000,000,000 times that of the sun, or $2.8 \times 10^{38}$ tonnes. The galaxy may contain as many as 300,000,000,000 stars, most of which, of course, are less massive than the sun.

By 1920, thanks chiefly to the work of the American astronomer Heber Doust Curtis (1872–1942), it was known that our galaxy was not the only system of stars in existence. There were others, stretching outward as far as the telescope could penetrate. Again, most of these other galaxies were smaller than ours, some with masses only a few hundred million times that of our sun, but there were some that were larger, with a few as massive as ten trillion times the mass of the sun.

There are some estimates that there are about 100,000,000,000 galaxies altogether, and that the total mass of the universe may be therefore something like $2 \times 10^{49}$ tonnes—that is, 20,000,000,000,-000,000,000,000,000,000,000,000,000,000,000 tonnes.

It is amazing that from our place on a small planet of an average star in the outskirts of a particular galaxy, we could, less than two centuries after determining the mass of Earth, determine the mass of the entire universe. Yet more surprising still, perhaps, is the human penetration in the other direction—toward the ultimately small.


## DOWN TO CELLS

Human beings have always been aware of living things less massive than they themselves: dogs, cats, barnyard fowls, and innumerable wild animals.

The usual example of a very small living creature, however, is the mouse. "As small as a mouse" is a common cliché.

The common house mouse is indeed a small animal, but there are many species of mice, and the house mouse is not the smallest. That distinction goes to the European harvest mouse, which is only half the size of the house mouse. The harvest mouse, as an adult, may weigh as little as 5 grams.* The average American male human being, then, has a mass about 14,600 times that of a harvest mouse.

Yet the harvest mouse is not the smallest mammal. That mark belongs to the pygmy shrew, which weighs about 2 grams. A pygmy shrew is to a man what a man is to a giant sequoia tree.

The smallest bird is the bee hummingbird, which is also 2 grams in weight as an adult.

It isn't possible for a warm-blooded creature, such as a mammal or a bird, to be any less massive than 2 grams.

As any object decreases in size its volume goes down considerably faster than its surface area does. This means that a small animal has more surface for its volume than a large animal of the same general shape has. An animal's body-heat production is related to its volume; its loss of heat is related to its surface. As an animal decreases in size, therefore, it loses heat at a faster and faster rate compared with the amount of heat it generates.

By the time an animal is as small as a shrew or hummingbird, it can just barely produce heat at the same rate it loses it, provided it eats virtually all the time. Shrews and hummingbirds therefore eat just about constantly through all their waking hours, and even so they are constantly hungry.

Cold-blooded animals, however, are not limited by heat loss, and can be smaller than any warm-blooded animal. Thus, whereas a harvest mouse is about 70 millimeters † long, including its tail, the

---

* A gram is equal to 1/1,000th of a kilogram, or (in common measure) of about 1/28th of an ounce.

† A millimeter is 1/10th of a centimeter, or 1/1000th of a meter, or (in common measure) 1/25th of an inch.

smallest lizard is only about 40 millimeters long, including its tail, and there is a frog that is only about 10 millimeters long.

The smallest vertebrate of any kind is a tiny fish called the pygmy goby. It is only about 8 millimeters long and has a mass of only about 4 or 5 milligrams.* A pygmy shrew has a mass equal to that of 450 pygmy gobies.

By and large, insects are smaller than vertebrates.

There is, to be sure, some overlapping. The largest insect, the Goliath beetle, can grow to be nearly 15 centimeters in length and can have a mass of as much as 100 grams. It is fifty times as massive as a pygmy shrew.

This is most exceptional, however. Flies and mosquitoes are more characteristic where insect sizes are concerned. The smallest insects are so-called fairy flies, which are only 2/5 of a millimeter in length. These are so small that they can be seen at best as a mere speck in a strong light.

When we get down to organisms that small, it might seem reasonable to ask if they can really be treated as alive. That anything smaller still should be treated as alive might seem out of the question.

Prior to modern times, no one thought of wondering if there might be anything alive that was so small as to be invisible to the unaided eyes. Tales of invisibility were told in plenty, but always the invisible objects were of quite perceptible size—even large—and were invisible only through magic spells.

Since early times, it must have been noted that glass objects sometimes made things they rested on seem larger, though distortions usually robbed the magnification of any use. It was not till the 1400s that lenses were used that made it possible to study small objects, such as insects, more easily than by the unaided eye.

The problem was to grind the lenses out of sufficiently high-quality glass and in a sufficiently smooth and accurate curve to reduce distortion and thus increase the useful level of magnification.

A Dutch microscopist, Anton van Leeuwenhoek (1632–1723), was the first who made the small magnifying lens useful enough to be called a "microscope." He used his lenses to look at everything, and in 1677 he found, in ditch water, small creatures that showed every sign of being alive, though they were too small to be seen unless they were magnified. He had discovered "microorganisms."

Making use of a "compound microscope" (one with more than a single lens), the English physicist Robert Hooke (1635–1703) had

* A milligram is equal to 1/1000th of a gram.

discovered even earlier, in 1665, that cork was made up of tiny rectangular holes too small to be seen without a microscope. He called them "cells."

It was occasionally observed at later times that tissues were divided into small segments, which, in plants at least, were marked off by distinct woody partitions. Cork, being dead tissue, consisted of the woody partitions with the interiors empty. The segments were still called cells even when not empty, and the partitions were called "cell walls."

The German botanist Matthias Jakob Schleiden (1804–81) suggested, in 1838, that all plants were composed of cells, these cells being the units of structure of plant tissue. The German physiologist Theodor Schwann (1810–82) suggested the following year that the same was true of animals, except that in animals the cells are separated by thin membranes rather than by comparatively thick walls and so are less noticeable. Between the two, they had worked out the "cell theory."

All organisms seen by the naked eye, from the largest sequoia tree to the smallest fairy fly, are made up of a number of cells, very often a vast number. They are "multicellular organisms." The average adult human being is made up of about 50,000,000,000,000 cells.

On the other hand, the microorganisms that Leeuwenhoek had discovered were made up of single cells and were "unicellular organisms." Naturally, unicellular organisms are far smaller than multicellular organisms, but they are just as alive.

Yet, to be sure, some one-celled objects are surprisingly large. Eggs, for instance, are one-celled objects. The cell, proper, is the yolk of the egg, the white and the shell being "extracellular." As for the yolk, that is almost entirely food intended for the developing embryo. The actual living substance is a mere microscopic spot on the surface of the yolk.

Eggs laid outside the body for independent development are particularly large because the food supply within is all the developing embryo will have until it is large enough to live outside the egg. The most familiar eggs are those of birds, and, generally, the larger the bird, the larger the egg.

A small hummingbird will lay an egg that is about 1 1/4 centimeters long, and that has a volume of about 0.4 cubic centimeters.* A hen's egg, the familiar breakfast egg so many of us eat, is about 6 1/4 centimeters long and has a volume of 50 cubic centimeters. A hen's egg is thus 125 times as voluminous as a hummingbird's egg.

---

* A cubic centimeter is a little over 1/16 of a cubic inch.

The largest living bird is the ostrich. The height of the back of a male ostrich is as much as 1.4 meters, and its long neck can hold its head up to 2.7 meters above the ground, half again as high as the head of even a tall human being. Masses of as much as 156 kilograms, twice that of the average human male, have been reported. Naturally, the ostrich lays an egg that is larger than that of any other bird.

The ostrich egg measures about 20 centimeters in length and 15 centimeters in width, with a mass of up to 8.5 kilograms. Its volume is as much as 1,200 cubic centimeters, so that an ostrich egg is about twenty-four times as voluminous as a hen's egg.

The ostrich egg is not, however, the largest egg in existence. Fish also lay eggs, and the largest living fish is the whale shark, which can have a length of 18 meters and a mass of 40 tonnes, making it about a third as massive as the largest whales.

It does not lay an ovoid egg as birds do, but lays an egg case that is irregularly shaped. One was found which was 30 centimeters long, 13 3/4 centimeters wide, and 8 3/4 centimeters thick. The volume was something like 3,600 cubic centimeters, or three times the volume of an ostrich egg. (We should remember, however, that a whale shark is about 250 times as massive as an ostrich.)

If we include extinct creatures, then dinosaurs do better still. The largest dinosaur eggs ever found were 30 centimeters long and 25 centimeters wide. The volume of such an egg would be about 5,000 cubic centimeters, or a little over four times that of an ostrich egg.

There are also extinct birds. The tallest bird that has ever existed was the giant moa of New Zealand, which attained a height of up to 4 meters (twice the height of a very tall man) and had a mass of up to 225 kilograms, nearly half again as much as an ostrich.

Not quite as tall but more massive is the aepyornis ("tall bird"), also known as the "elephant bird." Its height was some 3 meters at most, but its mass reached a maximum of 450 kilograms, nearly three times that of an ostrich.

The aepyornis egg was up to 37.5 centimeters long and had a volume of 8,900 cubic centimeters (2 1/3 *gallons,* in common measure). The aepyornis egg was eight times as voluminous as an ostrich egg, 180 times as voluminous as a hen's egg, and 22,500 times as voluminous as a hummingbird egg. Yet even the aepyornis egg was a single cell, the largest single cell that has ever existed on Earth.

The eggs of mammals are much smaller. These develop inside the mother's body and the developing embryo is fed primarily by way of a placenta, across which nutrients from the mother's bloodstream can diffuse into that of the embryo, while wastes diffuse in the other direction. The mammalian egg cell, or "ovum," does not, therefore, need the enormous food supply found in the eggs of birds, reptiles, and fish.

The human ovum, for instance, is pin-head size. It is only 1/280,000 the size of even a hummingbird egg and less than 1/6,000,000,000 the size of an aepyornis egg.

It is no use measuring the volume of the human ovum in cubic centimeters. We would have to work with inconveniently tiny fractions. For this, and for other cells, it would be better to use cubic micrometers.* The human ovum has a volume of about 1,400,000 cubic micrometers.

The human ovum is not quite as large as some microorganisms. The amoeba, for instance, shares many properties in common with animals generally, and is a member of a group called "protozoa" ("first animals"). The amoeba consists of a single cell, but since it lives independently, that one cell must perform all the essential functions of an animal. It must, therefore, be moderately large. The volume of an amoeba is about 4,200,000 cubic micrometers, so it is three times as voluminous as the human ovum.

There are some protozoa which are larger still, large enough indeed to be visible to the naked eye and to be larger than various small multicellular animals. The largest existing protozoan can be up to 1.5 centimeters across. There are extinct protozoa, called Nummulites, that were 2.4 centimeters across. Such a huge one-celled organism may well have rivaled a pygmy shrew or a bee hummingbird in mass.

Most cells are far smaller. A mammalian ovum must contain some food to keep the developing embryo going during the very early stages before the placenta has quite established itself. Therefore, the human ovum, small though it is compared to the amoeba, is still the largest cell in the human body.

The small size of cells making up a multicellular organism, as compared with those of many unicellular organisms, should not be surprising. The functions of a multicellular organism are distributed among the various types of cells that make it up. Each type of cell specializes in one function or another and need not possess the capacity for all as the one cell of a unicellular organism must. A body cell can therefore be considerably smaller than an amoeba and still do, efficiently, what it has to do.

The human liver cell, for instance, which is a busy chemical fac-

* A micrometer is 1/1,000,000th of a meter, or 1/10,000th of a centimeter. A cubic micrometer is, therefore, 1/10,000th × 1/10,000th × 1/10,000th, or 1/1,000,000,000,000,000th (one trillionth) of a cubic centimeter.

tory, has a volume of only 1,750 cubic micrometers, only 1/800 the volume of a human ovum and 1/2,400 the volume of an amoeba.

## DOWN TO BACTERIA

The cell itself does not represent the horizon of smallness for living things. Cells are not just homogeneous blobs of living matter. If they are studied under the microscope, they can clearly be seen to consist of smaller objects of considerable variety, each, no doubt, with its function.

The most important of these subcellular objects was first described in 1831 by the Scottish botanist Robert Brown (1773–1858). Because it was located near the center of the cells he studied, he called it the "nucleus" ("little nut"); it resembled a nut inside a cellular shell.

It was eventually discovered that the nucleus of a cell is particularly important with respect to reproduction, and to the manner in which a cell divides in two in such a way as to give each daughter cell the full properties and capacities of the mother cell.

In the blood there are vast numbers of cell-like objects that are not true cells because they do not ever divide and therefore do not require nuclei. They are manufactured in different places in the body, chiefly in bone marrow, out of cells that originally *do* have nuclei. The final product is specialized for one function: the combining with oxygen at the lungs, and then the releasing of that oxygen to body cells generally. These objects are "red blood corpuscles."

Red blood corpuscles don't need much volume to perform their function. The volume of the human red blood corpuscle is about 90 cubic micrometers, not quite 1/20 the volume of a liver cell.

The blood also contains still smaller objects, "platelets," which may have a volume of only 7 cubic micrometers. They are involved with blood-clotting. The platelets, like the red blood corpuscles, lack nuclei and are not true cells. They are the smallest cell-like objects in the body.

As for the size of a nucleus, it varies according to the type of cell, but a typical volume would be 35 cubic micrometers, thus they are smaller than a red blood corpuscle but larger than a platelet.

Nor is the nucleus itself a homogeneous mass of living matter. In 1882, the German anatomist Walther Flemming (1843–1905) described in detail the events that took place within a nucleus during cell division. He made these events clearer by using a red dye that would color some of the objects within the nucleus and not others.

The objects that received the dye he called "chromosomes" ("colored bodies"), and it was clear that it was the chromosomes that were most involved in cell division. Each chromosome split in two prior to the division, and then all the chromosomes separated so that each daughter cell got a complete set.

The chromosomes exist in pairs in each cell. When the female organism forms egg cells, each egg gets only a half-set of chromosomes, one of each pair. When a male organism forms spermatozoa (or sperm cells, the male analog of egg cells), each one of those gets only a half-set of chromosomes, too.

When organisms reproduce sexually, a sperm cell of the male combines with an egg cell of the female. The "fertilized ovum" that results now has a half-set of chromosomes from the mother and a half-set from the father. The two, in combination, form a full set, and the fertilized ovum develops into an organism which inherits characteristics from its father and its mother equally.

The difference between the egg cell and the sperm cell is that the former contains a half-set of chromosomes, plus a supply of food that must endure till the placenta is formed. The sperm cell has only a half-set of chromosomes and very little else. Its only task is to make use of its tail (for a sperm cell looks like a very tiny tadpole) to get that half-set it carries to the ovum. Once the sperm enters the ovum, that's it.

The sperm cell is therefore a very small cell, for it is made up of little more than a semi-nucleus, though that is enough to make it a true cell. The human sperm cell has a volume of only about 17 cubic micrometers, less than a fifth that of the red blood cell. It is larger than the platelet, but the sperm cell is the smallest true cell in the human body, just as the egg cell is the largest.

The human egg cell is about 82,000 times as large as the human sperm cell. To a sperm cell, an egg cell would appear forty times as large as a blue whale would appear to us.

Are sperm cells and platelets the ultimate horizon of smallness for living things? No, not at all!

In 1683, van Leeuwenhoek, looking through one of his small single-lens microscopes, detected objects just at the limit of his vision. He described them and drew pictures of them. He didn't know what they were, but looking at what he drew, later microbiologists knew he had seen what came to be called "bacteria." No one else saw them for over a century, till the Danish biologist Otto Frederik Müller (1730–84) made them out in somewhat better detail in a somewhat better microscope.

The difficulty with seeing bacteria was that the lenses of microscopes formed rainbowlike spectra, in addition to magnifying small objects. This meant that every object seen through them was surrounded by a colored halo. The smaller the object, the more pronounced the halo in comparison, and bacteria had so pronounced a halo that it was difficult to make them out at all.

Finally, in 1830, an English optician, Joseph Jackson Lister (1786–1869), devised lenses that would not form spectra and that would allow small objects to be seen without colored blurring. It was from that time only that bacteria could really be studied.

The push to do so was lacking, however, since there seemed to be no great importance to these tiny bits of life—at least until the French chemist Louis Pasteur (1822–95) got to work. In 1865, he was investigating a disease of silkworms that was threatening to wipe out France's silk industry. He located tiny bacteria in the diseased silkworms that were not present in the healthy ones, and it occurred to him that such small parasites could account for infectious disease, and that to prevent or cure such disease meant preventing the parasites from taking hold, or killing them once they had.

This "germ theory of disease" (" germ" being a name for any tiny bit of life, such as a bacterium) was probably the most important single advance in the history of medicine, and led the way to the doubling of the life expectancy of human beings within the next century. It also led to a new and profound interest in bacteria.

The German botanist Ferdinand Julius Cohn (1828–98) threw himself into the work and was the first to treat "bacteriology" as a special branch of knowledge. In 1872, he published a three-volume treatise on bacteria which may be said to have founded the science.

The study of bacteria pushed life to new horizons of minuteness. The largest bacterium known may have a volume no greater than 7 cubic micrometers, making it no larger than a platelet and less than half the size of a human sperm cell.

As for the smallest known bacteria, they are tiny bits of life indeed. Bacteria known as "pleuro-pneumonia-like organisms" (PPLO) were discovered in sewage in 1936. Each PPLO cell is a tiny sphere only 0.1 micrometers in diameter. That gives it a volume of 0.005 cubic micrometers.

This means that the smallest bacterium is only 1/1,400 as voluminous as the largest and only 1/3,500 as large as a human sperm cell.

The PPLO cell is the smallest free-living organism in existence, as far as we know—the most minute bit of life capable of living and reproducing on its own (provided, of course, the cell has a suitable environment, including food). PPLO cells are the tiniest cells which contain all the chemical machinery needed for life.

It is estimated that these smallest bits of free life have a mass of something like $10^{-15}$ gram (a quadrillionth of a gram). It would take

73,000,000,000,000,000,000,000 of them to weigh as much as the average American male.

## DOWN TO VIRUSES

And yet, unlikely though it may seem, even the PPLO cell does not represent the smallest form of life.

There are some infectious diseases for which scientists could not find some associated microscopic parasite. Either the germ theory of disease was wrong, or the parasites were there but were not seen. One reason why they might not be seen was that they were too small to see, even with the best microscopes of the day. Pasteur himself believed the "too-small" hypothesis.

In 1892, a Russian botanist, Dmitri I. Ivanovski (1864–1920), investigated "tobacco mosaic disease," a disease of the tobacco plant, with symptoms that included the mottling of its leaves. If a diseased leaf is mashed and the juice is placed on a healthy plant, the healthy plant catches the disease.

Ivanovski forced the juice of an infected leaf through a fine filter, one that was fine enough to filter out even the smallest bacteria—and yet the juice was still infectious. In 1895, this was repeated independently by a Dutch botanist, Martinus Willem Beijerinck (1851–1931). The latter called the infectious agent a "virus," a Latin word meaning "poison," since he didn't know what else to call it.

The British bacteriologist William Joseph Elford (1900–52) used a still finer filter in 1931, and did manage to stop the infectious agent. This showed that the virus was a particle (presumably living) that was somewhat smaller than the smallest bacteria.

The American biochemist Wendell Meredith Stanley (1904–71), working with a great quantity of juice from diseased tobacco leaves, managed to get rid of everything but the virus, and then, in 1935, crystallized it. He obtained a mass of fine needlelike crystals, which he isolated and found to possess all the infective properties of the virus in high concentration.

Meanwhile, a new kind of microscope was developed. Ordinary microscopes magnify objects by bending the light reflected from them. This will work only if the objects being magnified are at least somewhat larger than the light waves themselves. If not, the light waves step over them, so to speak. Light waves are very small, but when you get down to the size of the smallest bacteria, light waves are on the edge of being too large to work with.

Electrons are very tiny particles that we'll discuss later on. In 1927, the American physicist Clinton Joseph Davisson (1881–1958) showed that the electrons, which usually behaved as though they were particles, also behaved in some ways as though they were waves. These waves were much shorter than light waves, and could therefore be used to view objects much smaller than those that would be made visible by the use of light.

Electron waves are not part of the electromagnetic spectrum, and they can't be treated in the same way. Light waves can be bent by the use of lenses, while electron waves, being associated with an electrically charged particle, can be bent by the use of electromagnetic fields. Making use of this principle, the Russian-American physicist Vladimir Kosma Zworykin (1889– ) worked out an "electron microscope," and by 1939 he had an instrument which could penetrate into the depths of smallness fifty times farther than the best optical instrument.

The electron microscope made it possible to study viruses as visible objects for the first time, but there was, to begin with, considerable controversy as to whether they could be considered alive.

It was clear that viruses did *not* have all the chemical machinery needed for independent life. They could, however, manage to penetrate a cell, and within the cell they made use of the cell's machinery for their own purposes. They could multiply within the cell, at the expense of the cell, sometimes killing the cell in the process. It was a deeper form of parasitism than biologists had yet encountered, but could it be considered life?

The fact that Stanley had crystallized viruses seemed an argument against their being alive, since crystals were associated with chemical substances and not with living organisms. However, viruses were so small that they could be viewed as living organisms that, in some ways, acted as chemical substances.

The crucial decision came when scientists came to recognize the key role performed in all living organisms by nucleic acids, including a variety known as "deoxyribonucleic acid," which is usually abbreviated as "DNA." This was first understood as a result of the work of a Canadian physician, Oswald Theodore Avery (1877–1955), in 1944, and of the combined work of an English biochemist, Francis H. C. Crick (1916– ), and an American biochemist, James Dewey Watson (1928– ), in 1953.

Crick and Watson showed that DNA is the essential class of molecule in chromosomes and that it controls the manufacture of proteins in cells and tissues. It also controls the manner in which characteristics are inherited by daughter cells from mother cells in cell division, and by children from parents in the reproduction of organisms. The unit of the chromosome which controls the manufacture of a single variety of protein is called a "gene."

It turned out that all the viruses consist of nucleic acids wrapped in a protein shell. The larger, more complex ones have DNA as the variety of nucleic acid they possess. The smaller, simpler ones have a slightly different variety, usually abbreviated as "RNA." Tobacco mosaic virus contains RNA, for instance.

A virus might be looked on as a gene, or a group of genes, existing in isolation, capable of penetrating a cell and taking over (as invaders) the control ordinarily exercised by the cell's own genes.

In considering the volume of a virus particle, it would be better to use a new unit, the "cubic nanometer," where 1 cubic nanometer is equal to a billionth of a cubic micrometer. For instance, the PPLO cell, which has a volume of 0.005 cubic micrometers, can also be said to have a volume of 5,000,000 cubic nanometers.

There are some viruses that are so large they overlap the bacteria range. There is a group of particularly large viruses, for instance, that are called "rickettsia," because they were first identified by the American physician Howard Taylor Ricketts (1871–1910), in 1906. They are large enough to be seen under the optical microscope and can have as much as ten times the volume of the smallest free-living bacteria. The rickettsia, however, are lacking in at least one vital chemical component and can only multiply within cells, as is true of viruses.

The rickettsia cause diseases like Rocky Mountain spotted fever and typhus fever. The volume of the typhus fever rickettsia is 54,-000,000 cubic nanometers.

A smaller virus, like that which causes influenza, has a volume of 800,000 cubic nanometers and is only a little over 1/6 the size of the smallest bacterium. The tobacco mosaic virus is 50,000 cubic nanometers in volume, only 1/100 the size of the smallest bacterium (no wonder it passed through Ivanovski's filters), and the average gene has a volume of 40,000 cubic nanometers.

The smallest known virus is that which causes spindle tuber disease in potatoes. It may have a volume of only 200 cubic nanometers. At that volume, it would be about 1/2,500 the size of the smallest bacterium, and would have a mass of $8 \times 10^{-19}$ grams.

There is an amazing spread in life forms on Earth. It would take $2.5 \times 10^{27}$ spindle tuber disease viruses (two and a half billion billion billion) to have the mass of a sequoia tree. And yet the tiny virus is as alive as the massive sequoia tree.

Nevertheless, even the smallest virus is not the ultimate; it is made up of objects that are smaller still, and by the time human beings were able to study viruses and determine their size, the horizon of smallness had pushed far beyond them.

# 19
# Atoms and Beyond

## THE DIVISIBILITY OF MATTER

The ancient Greeks argued over the extent to which any object could be divided. Pottery could easily be cracked, and the broken shards cracked again into smaller shards, until it was all beaten into a fine dust. Could the dust be beaten finer and finer without limit? Or was there such a thing as a piece of pottery (or of anything else) so small that it could not be divided, by any means, into anything smaller?

Neither those philosophers who thought there was such a thing as a smallest object nor those who thought there was not had any experimental observations on their side. Both argued logically from basic assumptions.

Naturally, though, if one starts with different assumptions, one ends with different conclusions. Since neither set of conclusions is checked by comparison with the universe itself, there is no clear way of choosing between them. One chooses the alternative that makes one feel more comfortable.

On the whole, the most influential of the Greek philosophers chose indefinite divisibility and the absence of any ultimate particle.

The first person we know of by name who upheld the minority notion of an ultimate particle was the Greek philosopher Leucippus. He maintained this about 450 B.C. He also seems to have been the first to state "the rule of causality," which states that every event has a natural cause. This eliminates any consideration of the supernatural, or of magic.

Better known than Leucippus is his pupil Democritus (470–380 B.C.). Democritus maintained that all matter consisted of tiny particles so infinitesimally small that matter *seemed* continuous. He felt these particles to be so small that nothing smaller was conceivable. Because he considered these particles to be indivisible, he called them *atomos,* the Greek word for "indivisible," and this becomes "atom" in English.

The atoms, Democritus held, were eternal, unchangeable, and indestructible, and except for them there was nothing. They existed in different varieties, and all the different objects that exist are made up of atom combinations. If the objects we know are different, it is only because the atom combinations that make them up are different.

In all this, Democritus was essentially correct, in the light of modern thought, but he had no way of demonstrating that he was correct, and his views were dismissed.

Democritus is supposed to have written seventy-two volumes altogether, but his views were not so popular that scribes were encouraged to copy those books many times over. The fact that relatively few copies were prepared in the first place is probably the chief reason why no copies exist today. We know of Democritus' views only because they were referred to (usually adversely) by other philosophers whose works do survive, at least in part.

Nevertheless, the "atomism" of Leucippus and Democritus never entirely vanished. There were always a few philosophers who accepted the minority view.

The most important of the later Greek philosophers to do so was Epicurus (341–270 B.C.), who founded a popular philosophic school, and a course of philosophy that came to be known as "Epicureanism." He taught an atomist view of the universe, and this was carried on by his pupils. Not one of Epicurus' 300 books survives today.

The Roman writer Titus Lucretius Carus (95–55 B.C.) was an Epicurean. In 56 B.C. he published *De Natura Rerum* ("On the Nature of Things"), in which he described the universe in thoroughgoing atomistic terms. He felt that even the mind and soul were made up of atoms, and that if there were any gods, they were made up of atoms, too.

Lucretius' one book did not survive either. At least, there were no copies known to be in existence during the Middle Ages, and any knowledge of the book was from references in other sources. But then, in 1417, a single manuscript was discovered, quite unexpectedly, and copied. It proved popular.

In 1454, the German inventor Johann Gutenberg (1398–1468) devised the art of printing with movable type, and it became so easy to produce identical copies of any book in great numbers that from that day to this, no book of any importance whatsoever has been totally lost.

One of the earliest books to be printed was Lucretius' book, and so Greek notions of atomism survived after all.

One of the moderns who was influenced by Lucretius' book, and who adopted the notions of Greek atomism, was the French philosopher Pierre Gassendi (1592–1655). Gassendi's writings, in turn, influenced an Irish physicist and chemist, Robert Boyle (1627–91). It was Boyle, finally (twenty-one centuries after Leucippus had come up with the notion, to begin with), who considered atomism in the light of experimentation.

He experimented with air. Air is much less dense than liquids or solids, such as water or rock. That is, a given volume of air has a mass that is roughly 1/750 that of the same volume of water, and less than 1/2,000 that of the same volume of rock.

This difference in density could arise out of one of two possibilities, from the atomistic standpoint. Either the atoms of air are themselves much less dense than those of water or rock, or the atoms of air are fully as dense as those of water or rock but are spread out more thinly (or, to some extent, both, of course).

Suppose the case is that the atoms of liquids and solids are in contact with each other, while the atoms of air are far apart and are separated by nothingness, and that it is this which makes air so much less dense than other substances. In that case, the atoms of water or rock, being in contact, or nearly in contact, should be difficult to press together (to "compress") and be made to take up less volume. Air, on the other hand, should be easily compressed, as the separated atoms are forced more closely together. Air could be squeezed together in the same way a sponge could be squeezed together, and for the same reason. (This was something which had been suspected by an ingenious Greek engineer, Hero, about A.D. 60.)

In 1662, Boyle tested this. He made use of a J-shaped glass tube, with the short end closed. He poured mercury into the long, open-ended arm, and it collected at the bottom of the J, trapping air in the short end. As he poured in more mercury, the additional weight of the mercury compressed the air in the short arm. This was not true for liquids and solids, and this was a powerful blow in favor of atomism.

One might argue, of course, that this was only evidence for the atomistic makeup of air and other gases. Liquids and solids might still be continuous. However, water can easily be made to boil, when heated, or to evaporate at ordinary temperatures, and, in either case, it becomes water vapor, which is a gas, and can be easily compressed. Many substances that are solid or liquid can be converted into vapor form, and so to accept the fact that gases consist of atoms means accepting the fact that *all* substances probably consist of atoms.

It might be, of course, that substances do break up into tiny particles, but that the process is random so that the particles are of any size and have no fundamental significance. Evaporation and boiling might merely break up a substance, as hammering breaks up rock, and the particles formed might always break up into still smaller pieces.

This notion was put to rest beginning with the work of a French chemist, Joseph Louis Proust (1754–1826). In 1799, Proust showed that a substance, copper carbonate, was made up of copper, carbon, and oxygen, and that it was always made up of the same proportion by weight of each of these. No matter how copper carbonate might be prepared in the laboratory, or how it was isolated from the rocks, 10 grams of copper carbonate always contained 5 grams of copper, 4 grams of oxygen, and 1 gram of carbon.

He went on to show that a similar situation existed for a number of other substances, and formulated the generalization that all compounds (substances made up of different varieties of atoms) always contained elements (substances made up of a single variety of atoms) in certain definite proportions and no others. This is called "the law of definite proportions."

Now if matter were continuous and could break into pieces of any size, it would seem very likely that different elements could combine in any proportions, just as powdered sugar and powdered cocoa can be mixed in any proportions. To have a law of definite proportions would therefore seem to imply that each element was made up of certain fundamental particles of definite size, which could only combine in certain relative numbers. And those fundamental particles must be the atoms that Leucippus and Democritus had talked of.

## ATOMIC WEIGHTS

Making use of Proust's findings, and of other chemical observations that lent increasing credibility to the notion of atomism, the English chemist John Dalton (1766–1844), beginning in 1803, formulated an "atomic theory of matter." He indicated the debt owed to the ancient Greek philosophers by accepting the word "atom" for the ultimate particles.

He argued that all atoms of a particular element were identical, and that the atoms of one element were different from those of all other elements. So far he went along with Democritus. However, Dalton went further; he held that atoms differed in mass, and that their relative masses could be measured. He even advanced suggestions as to precisely how the masses of certain atoms were related to each other.

Dalton's figures for "atomic weights," as these relative masses were termed (and improperly so, for "atomic masses" would have been better), were not very accurate. The first reasonably accurate atomic weights were worked out, beginning in 1828, through the careful analyses of various compounds by the Swedish chemist Jöns Jakob Berzelius (1779–1848).

Berzelius used the mass of the oxygen atom as his standard, setting it equal to 16. With that as standard, the atomic weight of sulfur is 32. In other words, the sulfur atom is twice as massive as the oxygen atom.

There are atoms that are more massive still. The most massive atom that occurs in reasonable quantities in nature is that of uranium. Its atomic weight is 238, so that the uranium atom is nearly fifteen times as massive as the oxygen atom. Even more massive atoms have been manufactured in the laboratory in recent decades, with their atomic weights pushing the 260 mark.

In the other direction, there are atomic weights less than that of oxygen. The atomic weight of nitrogen is 14, of carbon 12. The lowest atomic weight of all is that of hydrogen, for the figure for that is 1.

But how large are the individual atoms in terms of ordinary units of mass or volume?

Knowing the *relative* mass is of no help. You might know that the sulfur atom is twice as massive as the oxygen atom, but that alone would not tell you how much mass either atom actually possessed as measured in grams, or how wide either atom was in meters.

A possible key to the solution arose in 1811 when an Italian physicist, Amadeo Avogadro (1776–1856), advanced arguments that led to the conclusion that similar volumes of different gases contained similar numbers of particles of matter. This would mean that if one gas was three times as dense as another, it was because the individual particles making up the first gas were three times as massive as those making up the other. "Avogadro's hypothesis" could thus be used to determine atomic weight from relative densities.

The hypothesis was ignored, at first, because it was misunderstood. Chemists did not grasp the fact that the particles making up gases were not necessarily single atoms. They might be groupings of two or more atoms held together more or less permanently, these groupings being called "molecules." Thus, oxygen gas consists of oxygen molecules, each of which is made up of a pair of oxygen atoms, while water vapor is made up of water molecules, consisting of three atoms each, two hydrogen atoms and one oxygen atom.

This means that although equal volumes of oxygen and of water vapor contain equal numbers of particles (molecules!), the total number of *atoms* in the oxygen volume is only two-thirds that in the water vapor volume.

It wasn't until 1860 that this was finally cleared up. In that year, at an international conference of chemists (the first such conference ever to be held), the Italian chemist Stanislao Cannizzaro (1826–1910), in a strong presentation, clearly explained the difference between atoms and molecules, and how to use Avogadro's hypothesis successfully.

Given the atomic weight of oxygen as 16, chemists began to speak of the "molecular weight" of oxygen as 32, since each molecule of oxygen is made up of two oxygen atoms. The molecular weight of water is 18, since each molecule of water is made up of an oxygen atom (atomic weight, 16) and of two hydrogen atoms (each with an atomic weight of 1).

One may not know how much mass an individual oxygen molecule possesses, but we know that a certain number (call it $N$) of oxygen molecules weighs 32 grams. That same number of water molecules would weigh 18 grams, since the relative mass of a water molecule to an oxygen molecule is 18 to 31. In fact, that same number of *any* molecule would have a mass of $x$ grams, if that molecule happened to have a molecular weight of $x$.

Since this is a consequence of Avogadro's hypothesis, $N$ is called "Avogadro's number."

Now the question is: What is the value of Avogadro's number? Once that was determined, we would know the mass of an individual molecule and of the individual atoms of which that molecule was composed.

That, however, is not an easy question to answer. All that chemists could be sure of at first was that Avogadro's number was very large.

## THE SIZE OF ATOMS

The road to the answer had opened in 1827, when Robert Brown (who was later to discover the nucleus of the cell) viewed a suspension of pollen grains in water under the microscope. He noted that the individual grains were moving about irregularly. This, he thought, was the result of the life hidden within the pollen grains. However, when he studied dye particles of the same size, also suspended in water, he found the same erratic motion present, despite the fact that the particles were certainly not living.

The phenomenon was referred to as "Brownian motion," and for some decades it remained a mystery.

In 1860, the Scottish mathematician James Clerk Maxwell (1831–79) made a thoroughgoing analysis of the properties of gases, on the assumption that they were made up of atoms or molecules moving rapidly and randomly in all directions. The assumption explained the properties of gases very well, and this "kinetic theory of gases" was quickly accepted.

It was clear that the atoms or molecules of liquids and solids could not have the untrammeled movement that they would have had in gases, but there was surely *some* motion. In liquids, the atoms or molecules would jiggle past each other like people in a crowd; in solids, they might vibrate in place like restless soldiers in formation.

Once that was understood, Brownian motion could be understood also. Any object suspended in a liquid would be jostled from all sides by the moving molecules in the liquid. On the whole, equal numbers would strike from all possible directions, and the net result would be a balance, so that the object in suspension would not move. To be sure, the laws of chance would ensure that there would always be a few more striking from one direction than from another at any given moment, but out of many trillions, a few extra this way or that would not matter.

The smaller a suspended object was, the fewer the number of molecules of the liquid that would strike altogether, so that the existence of a few more this way or that would bulk larger in proportion. With objects as small as pollen grains, a few more from one direction would push the pollen noticeably; then a few more from another direction would push it in a new way. Inevitably, there would follow the random jiggling of Brownian motion. Here was something visible that depended upon the motion of a few individual atoms or molecules.

A Swedish chemistry student, Theodor H. E. Svedberg (1884–1971), suggested this interpretation of Brownian motion in 1902. Then Einstein, in 1905, the same year in which he worked out the Special Theory of Relativity, published a mathematical analysis of Brownian motion on this basis. In the final equation, Avogadro's number was included. If all the other quantities in the equation were known, the value of Avogadro's number could be calculated.

The French physicist Jean Baptiste Perrin (1870–1942) set about conducting an experiment that would allow him to determine the value of the various quantities in Einstein's equation.

In 1908, he suspended particles of gum resin in water. If they were subject to gravity only, the particles would drop to the bottom of the tube. Brownian motion keeps the particles jiggling, however, and suspends them at different heights above the bottom.

According to Einstein's equation, the number of particles found at increasing heights above the bottom of the tube should fall off in a certain way, and Perrin found that the number *did* fall off, exactly as predicted. From his observations, he worked out all the values in Einstein's equation except for Avogadro's number. Once that was done, Avogadro's number could be calculated. Perrin did so, and was the first person to get a reasonable notion of the actual size of atoms and molecules.

The best value we have for Avogadro's number today is just about $6.022 \times 10^{23}$, or a little over six hundred billion trillion. It would take that many oxygen molecules to have a mass of 32 grams; that many water molecules to have a mass of 18 grams; and that many hydrogen molecules (made up of two hydrogen atoms each) to have a mass of 2 grams. That many individual hydrogen atoms, the least massive of all atoms, would have a mass of 1 gram.

A single atom of hydrogen would weigh $1/6.022 \times 10^{-23}$ grams, or $1.67 \times 10^{-24}$ grams, or a little over a trillionth of a trillionth of a gram. (Other atoms would weigh proportionately more, of course.)

It's no wonder, then, that it took so long to decide that matter was composed of atoms, let alone to find out how massive they were. The wonder is that it could be done at all.

Since 18 grams of water has a volume of 18 cubic centimeters and contains $6.022 \times 10^{23}$ water molecules, it is easy to calculate how much volume a single water molecule takes up and, by working with other substances as well, how much volume single atoms of various kinds possess.

Hydrogen atoms, for instance, if assumed to be spherical in shape, would have a diameter of $1.35 \times 10^{-10}$ meters, or a little over a ten-billionth of a meter. More massive atoms are a little bit larger, but even the most massive atom known is probably not wider than $8 \times 10^{-10}$ meters in diameter.

This means that about 405 hydrogen atoms can be squeezed into a cubic nanometer, and that the spindle tuber disease virus, the smallest known fragment of life, can contain perhaps 75,000 atoms.

# ELECTRONS

And yet, astonishingly enough, by the time the size and mass of the hydrogen atom was determined, it was no longer on the horizon of smallness. Scientists had moved far beyond.

The ancient Greeks had defined atoms as the smallest particles that could possibly exist, and modern chemists of the 1800s had gone along with that. Certainly, all the chemical findings, for a century after Dalton's presentation of the atomic theory, seemed to confirm it. The hydrogen atom, as the smallest atom, seemed to be the smallest bit of matter that could possibly exist.

The beginning of a new advance came out of experiments with electricity. Scientists knew that an electric current would travel easily along metals and other "conductors." They knew that such currents could (under sufficient force) leap across nonconductors. Currents could, for instance, leap across air, producing a spark and a crackle.

The logical question, then, was whether an electric current could be forced through a vacuum. As it happened, in 1855, a German inventor, Heinrich Geissler (1814–79), devised a new and better way of pumping the air out of glass tubes. In this way, he created better vacuums for scientists to work with.

Metal was sealed into "Geissler tubes" at two places, with a vacuum gap between. An electric current was then forced from one of the pieces of metal, the "cathode," across the gap, to the other piece, the "anode."

A German physicist, Julius Plücker (1801–68), experimenting in this fashion in 1858, noted a greenish fluorescent glow at the cathode. Another German physicist, Eugen Goldstein (1850–1931), decided that the fluorescence consisted of radiation emerging from the cathode and, in 1876, he termed the phenomenon "cathode rays."

There followed a two-decade dispute as to whether the cathode rays were a lightlike radiation, or whether they consisted of a stream of particles. In 1895, Perrin (who was later to be the first to measure the size of atoms and molecules) showed that when cathode rays bathed a cylinder, that cylinder gained a negative electric charge that grew larger with time. Lightlike radiations were not known to be capable of carrying an electric charge, so that made it look as though cathode rays consisted of particles.

The cathode-ray particles were viewed as the fundamental particles of electricity, and the Dutch physicist Hendrik Antoon Lorentz (1853–1928) suggested they be named "electrons."

Since they were electrically charged, electrons ought to be deflected in their flight by other electrically charged particles, and by magnets, as well. The English physicist Joseph John Thomson (1856–1940) managed both to demonstrate deflections and, from the extent of those deflections, to calculate the mass of the electron in 1897. That mass turned out to be incredibly tiny—only 1/1,837 of the mass of a hydrogen atom, or 9.1 × 10⁻²⁸ grams.

The electron was not involved with electric currents only. In 1902, the German physicist Philipp E. A. Lenard (1862–1947) began to study certain electrical effects produced in metal when light fell upon it. He discovered that light caused the metal atoms to eject electrons. This "photoelectric effect" made it seem rather likely that atoms contained electrons. What's more, since various metals all emitted identical electrons, as nearly as could be determined, it began to seem that electrons were a common component of all atoms.

For the first time, it was borne in upon scientists that although atoms were the smallest objects they had to deal with in ordinary chemical events, those atoms were nevertheless complex objects built up of still smaller entities. The electrons were the first "sub-atomic particles" discovered.

## THE ATOMIC NUCLEUS

Meanwhile, in 1896, radioactivity had been discovered. It was quickly found that the radiations given off by uranium were of three types. These were named "alpha rays," "beta rays," and "gamma rays," after the first three letters of the Greek alphabet. The gamma rays, it turned out, were lightlike in character, but with extremely short wavelengths. The beta rays consisted of streams of speeding electrons.

The alpha rays were something new, however. They were streams of particles that were much more massive than electrons, over 7,000 times as massive, in fact, and, therefore, four times as massive as the hydrogen atom. Despite this, the alpha rays seemed to be unusually small, since they could pass through thin layers of matter, something atoms could not do.

The British physicist Ernest Rutherford (1871–1937) bombarded matter with streams of alpha particles. Beginning in 1906, he found, for instance, that alpha particles could pass through a sheet of gold foil half a micrometer thick as though there were nothing there. They were neither stopped, slowed, nor diverted from their path. To be sure, half a micrometer is exceedingly thin, but that is enough thickness to hold six or seven gold atoms. The alpha particles had to pass through them all.

But then, a few alpha particles *were* diverted from their paths—even quite sharply. In fact, a very few alpha particles would bounce directly backward when they struck the gold foil.

In 1911, Rutherford announced his interpretation of these observations. An atom, he said, consisted of a very small nucleus that contained almost all the mass of the atom, and around it was a cloud of very light electrons. The alpha particles, flashing through the gold foil, passed, for the most part, through the electron-filled outskirts of the atoms, since the outer layers of the atom took up almost all the volume of the atom. In doing this, they were not affected in any detectable way, since electrons are so much less massive than alpha particles are.

A few particles, however, by sheer chance, would just happen to pass near the massive nucleus of one of the atoms and would be deflected. A few alpha particles would actually strike a nucleus head-on, and would bounce back.

As it turned out, the nucleus carries a positive electric charge. In 1914, the English physicist Henry G. J. Moseley (1887-1915) showed that each different element has an atomic nucleus with a particular size of electric charge. In each case, the charge is neutralized by the negatively charged electrons in the outskirts.

Thus, an oxygen atom has a nucleus with a charge of +8, while outside its nucleus are eight electrons, each with a charge of −1. As a result, the oxygen atom, as a whole, is electrically neutral. The uranium atom has a nucleus with a charge of +92, and possesses ninety-two electrons on the outskirts as a balance.

Rutherford suggested, also in 1914, that the atomic nucleus derived its positive electric charge from its possession of a number of particles, each of which possessed a charge of +1. This particle, with a unit positive charge, he called a "proton." The charge on the proton is precisely equal in size to the charge on the electron, though it is opposite in sign; yet the proton is 1,836 times as massive as the electron.

The nucleus does not contain protons only. In 1932, the English physicist James Chadwick (1891-1974) detected a subatomic particle that was just a trifle more massive than the proton (and that had 1,838 times the mass of the electron) but that carried no charge at all. It was electrically neutral and was named the "neutron."

The atomic nucleus is built up of protons and neutrons, and each different type of atom is built up of a different combination.

Each element has a fixed number of protons in its nucleus, but the number of neutrons can vary to a small degree. Each different number of neutrons results in a slightly different variety ("isotope") of the element. Thus, all oxygen atoms have eight protons in their nuclei. Most have eight neutrons as well, and they are then examples of the "oxygen-16" isotope. The 16 represents the total number of protons plus neutrons. A few oxygen atoms have nine, or even ten, neutrons in the nucleus in addition to the eight protons, and these are "oxygen-17" and "oxygen-18."

The smallest nuclei are those of hydrogen, for these carry a charge of $+1$. Almost all hydrogen atoms have nuclei that consist only of a proton, and nothing else, and these are "hydrogen-1." Hydrogen atoms may also have one neutron, or even two, in the nucleus, and these are "hydrogen-2"and "hydrogen-3."

As far as size is concerned, we begin to run into difficulties when we deal with subatomic particles. All particles have wave properties, and the less massive the particle is, the more pronounced the wave aspect. The electron has so little mass that it is difficult to speak of its having a particular size. It is more like a spread-out wave.

The protons and neutrons, however, are massive enough for the particle aspect of their nature to predominate, and, as particles, each is roughly $2.5 \times 10^{-15}$ meters in diameter or 2.5 quadrillionths of a meter. The more complex nuclei are somewhat larger than the smaller ones, as we pack more and more protons and neutrons into them. The largest naturally occurring nucleus is that of uranium-238, which contains 92 protons and 146 neutrons. Its diameter is $1.55 \times 10^{-14}$ meters, so that it is about 6.2 times as wide as a proton.

In any atom, the nucleus has about 1/100,000 the diameter of the atom of which it forms a part. If the atom were a hollow sphere, one could fill it with about $10^{15}$, or one quadrillion, nuclei.

## NEUTRINOS

Even the proton and neutron do not represent limits of smallness. In 1953, the American physicist Murray Gell-Mann (1929– ) suggested that the proton and neutron were made up of three still more fundamental particles he called "quarks." Despite this, we need not expect to get any further downward with respect to volume, for below the proton and neutron, volume ceases to have much meaning.

Where mass is concerned, however, it is the electron, with its mass of $9.1 \times 10^{-28}$ grams, that is closer to the ultimate horizon. The electron, in fact, is the least massive particle that is known to carry an electric charge, and there has long been the feeling that in this respect we have reached the horizon of smallness.

There are, indeed, particles with less mass than the electron, but they are all without electrical charge. What's more, these less-than-electron particles all have (or are suspected of having) zero mass, and so are not usually considered to be particles of matter.

There are three varieties of zero-mass particles: photons, which are the fundamental particles of light and related radiation; gravitons, which are thought to be the fundamental particles of gravitational interactions, but which haven't been detected yet; and neutrinos.

The neutrinos are the most nearly nothing of the three, it would seem. While photons and gravitons interact readily with matter, neutrinos do not. They pass through matter as though it were not there at all. A beam of neutrinos can pass through the entire sun and scarcely be disturbed in the process. An occasional neutrino out of many trillions may make a square hit on a nucleus and may then interact, but nothing more.

The existence of the neutrino was first postulated in 1931 by the Austrian physicist Wolfgang Pauli (1900-58) for purely theoretical reasons. The extreme difficulty of detecting something without mass, without charge, and without interacting tendencies was such that it wasn't till 1956 that it was detected by the American physicist Frederick Reines (1918- ).

In 1980, however, experiments both in the United States and in the Soviet Union have made it appear that, just possibly (for the observations are at the very limits of what can be detected), neutrinos have masses that are not *quite* zero. The mass of a neutrino may be about 1/13,000 that of an electron, or about 1/23,000,000 that of a proton.

If this is true, this places the neutrino at the lower horizon of mass, since its mass is something like $7 \times 10^{-32}$ grams, or seven hundred-millionths of a trillionth of a trillionth of a gram.

And now we can see the full range of mass in the universe. Suppose we imagine thirty giant sequoia trees. A neutrino is to them as they are to the entire universe out to the farthest star, and the human mind has stretched out in either direction to measure the mass of both the greatest and the tiniest.

# 20
# Density and Pressure

## THE ANCIENT DENSITIES

In dealing with mass, there are times when mere quantity is not all that counts. What may also be of importance is how tightly the mass is packed together—how much mass is to be found in a given volume. This "mass per volume" is "density," and a given substance, under a given set of conditions, has a characteristic density.

When the metric system was established, the value of the gram and the meter were deliberately chosen so that 1 cubic centimeter of water would have a mass of 1 gram. The density of water is therefore 1 gram per cubic centimeter.

Actually, the initial measurements were not quite accurate, and the density of water, under standard conditions of pressure and temperature, is now taken as 0.999973 grams per cubic centimeter. This is a deviation from 1 that is so small, of course, that it can scarcely be of interest to anyone but a professional scientist. The density of water also varies with pressure and temperature, but only to a relatively minor degree. We will continue to consider the density of water to be 1 gram per cubic centimeter.

The preferred system of units these days is the "SI version" (Système Internationale) of the metric system, in which kilograms and meters are used instead of grams and centimeters. A kilogram is, of course, equal to 1,000 grams, and a meter is equal to 100 centimeters.

A cubic meter is equal to 100 × 100 × 100, or 1,000,000, cubic centimeters. If a cubic centimeter of water weighs 1 gram, then

1,000,000 cubic centimeters (1 cubic meter) of water must weigh 1,000,000 grams, which is equal to 1,000 kilograms.

Water, therefore, has a density of 1,000 kilograms per cubic meter.

When water freezes, the molecules that make it up move into a looser configuration. The mass spreads out over a slightly larger volume and the density decreases. The density of ice is about 917 kilograms per cubic meter.

Solids float on water if they are less dense than water. Ice, therefore, floats on water. This is an unusual situation, since almost all liquids other than water grow denser when they freeze.

Another familiar substance that will float on water is wood. Different kinds of wood are made up of similar substances, which are themselves slightly denser than water. The wood fibers, however, are packed together more or less loosely and take up more volume than they would if they were compact.

In some kinds of wood, the fibers are indeed compact, and these are as dense as water or denser. Most kinds of wood are loosely packed, however, and have densities anywhere from half to three-fourths that of water. Balsa wood has its fibers very loosely packed indeed and has a density of only about 140 kilograms per cubic meter at most.

Commonly, it is said that wood floats on water because wood is "lighter" than water. This is an incomplete statement. What is really meant is that a *given volume* of wood is lighter, or less massive, than the same volume of water. In other words, wood is *less dense* than water, and that is why it floats.

Most common solid objects are denser than water and will, therefore, sink if placed in water. Rocks will sink, for instance. Different kinds of rock have different densities, but a typical density for the various rocks making up Earth's crust is about 2,800 kilograms per cubic meter, or 2.8 times the density of water.

Metals, in general, are denser than water. This rule has its exceptions. The metal lithium, for instance (the least dense of all the metals), has a density of only 534 kilograms per cubic meter and is therefore little more than half the density of water. (It can float in water, but it also reacts with water and gradually dissolves when in contact with it.)

Lithium, however, and all other not-very-dense metals were only discovered in modern times. The ancients knew of only seven metals (plus various mixtures, or alloys, of these), and all are considerably denser than water, or even rocks. It may well have been the unusual density, as well as the appearance, of small nuggets that first attracted the attention of early civilized human beings to the existence of metals.

The density of tin, for instance, the least dense of the seven metals known to the ancients, is 7,280 kilograms per cubic meter, which is 2.5 times the density of a rock such as granite. Among the other long-known metals, iron has a density of 7,860 kilograms per cubic meter; copper, one of 8,920 kilograms per cubic meter; and silver, one of 10,500 kilograms per cubic meter.

Denser than any of these, and yet common enough and cheap enough for ordinary people to have acquaintance with, is the long-known metal lead. It has a density of 11,300 kilograms per cubic meter and is therefore some four times as dense as rock.

It is not surprising, then, that lead has become a byword for "heaviness"; that we speak of "leaden spirits" when we are sad, of "leaden feet" when we are tired, of "leaden eyelids" when we are sleepy, and so on.

When objects of similar size are moving, the denser ones have more mass, and therefore more momentum, more kinetic energy, and greater shattering effect on collision. That is why missiles changed from rock to metal as military technology advanced, and why, in particular, bullets are made of lead.

Again, when surveyors want to set up a vertical line, they weight the line with a lump of dense material, often lead, to stretch it into straight verticality under the pull of gravity. Since the Latin word for lead is *plumbum*, we speak of a "plumb line."

And yet lead was not the densest material known to the ancients. Two known metals were denser, but they were rare, and ordinary people did not come in contact with them. They did not experience the extraordinary density they possessed, so lead remained proverbial.

One of these particularly dense metals is a liquid—mercury. That is astonishing, because most liquids are not particularly dense. Water, at 1,000 kilograms per cubic meter, is about the densest liquid known to the ancients (aside from mercury). Various plant and animal oils have densities of about 900 kilograms per cubic meter, alcohol has a density of 790 kilograms per cubic meter, and so on.

Mercury, however, has a density of 13,600 kilograms per cubic meter, and is therefore 20 percent denser than lead is. Someone who is unacquainted with the density of mercury and comes across a filled bottle of the metal in a chemistry laboratory and attempts to lift it with the sort of force that would suffice for an ordinary liquid of that volume could be forgiven if he thought, for a moment, that the bottle was glued to the table.

And even mercury does not hold the record. Gold has a density of 19,300 kilograms per cubic meter, and is, therefore, 70 percent denser than lead. Since gold is the least common and by far the most beautiful of the metals known to the ancients, it was not surprising to them that it represented an extreme.

Despite the fact that gold is so much denser than lead, gold's beauty predominated in the public mind, and no one uses it in an uncomplimentary way in a metaphor. You may drag on leaden feet, but you will dance on golden ones.

Why should solids and liquids differ so in density? Why should gold be thirty-six times as dense as lithium? In all of them, lithium as well as gold, the atoms are in contact, and differences in compactness of arrangement are rather minor.

Individual atoms, however, differ in mass. Some have many protons and neutrons packed into the nucleus and some have few. Atoms with many such nuclear particles have a higher atomic weight and are more massive. Consequently we would expect that, in general, the higher the atomic weight, the denser the element.

Lithium, for instance, has an atomic weight of 7, while oxygen and silicon (the chief components of rocks) have atomic weights of 16 and 28 respectively. Iron, on the other hand, has an atomic weight of 56, and gold, one of 197.

Nevertheless, the volume of the atoms and the manner in which they are packed do play a minor role. For instance, mercury has an atomic weight of 201 and lead has one of 207, yet, though each is a bit beyond gold in atomic weight, each is considerably below gold in density. The atom of highest atomic weight that occurs in nature in considerable quantity is uranium; and although its atomic weight is 238, which is 20 percent greater than that of gold, the density of uranium is about 19,000 kilograms per cubic meter, which is just a bit less than that of gold.

## THE MODERN DENSITIES

Gold may have first come to the attention of people about 4000 B.C., and for over five millennia it held the record for density. It would have been forgivable to feel the record would never be broken.

It was not until the 1740s that the record was, in fact, broken. In that decade, a Spanish scholar, Antonio de Ulloa (1716–95), studied nuggets of a metal that had been found in the sands of the Pinto River in Colombia. Since the metal was whitish, the Spaniards on the spot called it *platina del Pinto* ("little silver of Pinto").

Examination quickly showed that it was not silver. It was considerably denser than silver, considerably higher-melting, and considerably more inert (that is, considerably less likely to react with other substances). The new metal was therefore given a name of its own. The "-um" suffix usually used for metals was added to the Spanish name and it became "platinum."

Because of the high melting point and the inertness of platinum, it was excellent for use in the manufacture of chemical equipment and was much sought after for the purpose. About 1800, an English chemist, William Hyde Wollaston (1766-1828), devised a method (which he kept secret till nearly the end of his life) for working platinum in such a way as to make platinum vessels of high quality. The process made him rich, and that meant that other chemists studied platinum with even greater avidity.

In 1803, another English chemist, Smithson Tennant (1761-1815), discovered, in working with platinum, that two similar metals were mixed with it in small quantities. These he called "iridium" and "osmium"—iridium from the Greek word for "rainbow," because of the different colors of its compounds, and osmium from the Greek word for "smell," because its compound with oxygen had a foul stench.

It turned out eventually that the atomic weights of osmium, iridium, and platinum were, respectively, 190, 192, and 195, just below the atomic weight of gold.

In density, however, they were just above the mark of gold and, as it happens, in reverse order of the atomic weight. The densities are: platinum, 21,450 kilograms per cubic meter; iridium, 22,421 kilograms per cubic meter; and osmium, 22,480 kilograms per cubic meter.

Platinum is 11 percent denser than gold, and osmium is 5 percent denser than platinum. We now know enough about chemical substances to be quite certain that these three metals (and their alloys with each other) are the only materials that are denser than gold, under the conditions that prevail on the surface of Earth, and that, of them, osmium holds the record. We will find nothing that is denser than osmium.

Let us work, however, in the opposite direction. I have already mentioned lithium as the least dense metal, with its mark of 534 kilograms per cubic meter. This is indeed the least dense compact solid that exists under ordinary circumstances.

It is possible to have materials less dense, if they are not compact. Balsa wood is only about a quarter as dense as lithium, but balsa wood has its fibers so loosely arranged that much of its apparent volume is simply air. Its density, therefore, is an average of that of the actual wood and of the air it contains. Similarly, the overall density of a Ping-Pong ball is much less than that of balsa wood, because the ball's density is a weighted average of that of the thin spherical shell of celluloid and of the inner content of air. Soap bubbles are still less dense, because their shells of liquid are thinner and less dense than celluloid.

If we wish to confine ourselves to substances that are compact, then, in order to find anything less dense than lithium, we have to turn to materials that are only solids at low temperatures. (We will discuss temperature later in the book.)

The element with the lowest atomic weight of all, 1, is hydrogen. Hydrogen solidifies at a temperature of −260° C., and the solid hydrogen that then exists has a density of 86.6 kilograms per cubic meter. This is only 1/6 the density of lithium. Solid hydrogen is the least dense compact solid that can exist under any conditions, and osmium is 260 times as dense as it is.

Liquids, on the whole, are less dense than solids, and this is true in the case of hydrogen. Liquid hydrogen, at −253° C., has a density of 70 kilograms per cubic meter and is only 4/5 as dense as the solid form. It is the least dense liquid that can exist under any conditions.

That leaves us with gases.

Whereas in solids and liquids, the component atoms are in contact, in gases, the atoms (or molecules) of which they are composed are not. The ultimate particles of gases are separated by substantial volumes of nothingness (vacuum). Gases are therefore not compact, and their density is lower than that of liquids or solids for the same reason that the density of soap bubbles is low—because we are not dealing with a substance in bulk, but with a substance interspersed with a lower-density medium.

Nevertheless, we can define certain standard conditions of pressure and temperature, and measure the density under those conditions. This will give us standard densities we can deal with.

The atoms or molecules that make up gases are spread out equally (more or less) under given conditions. Consequently, the density is proportional to the mass of the individual atoms or molecules making up the gas.

Of the substances that are gases under standard conditions, that with the highest particle mass is radon. This is a radioactive material which breaks down quite rapidly so that it exists only in traces. If it could be collected in substantial quantities and its density measured, that density would turn out to be 10.2 kilograms per cubic meter.

Suppose, though, we consider uranium hexafluoride, which is a solid under ordinary conditions. It doesn't take much heat, however, to convert it into a gaseous vapor. The individual particles of such a gas are the molecules of uranium hexafluoride, each of which contains one uranium atom and six fluorine atoms. The molecular weight is 352 and the density of such a gas is about 16 kilograms per cubic meter.

Uranium hexafluoride vapors are probably the densest gas we can expect under conditions approaching the standard here on Earth, and yet it is less than a quarter as dense as liquid hydrogen, the least dense of all substances that are liquid or solid.

Air itself is composed of a mixture of oxygen and nitrogen. The density of oxygen under standard conditions is 1.43 kilograms per cubic meter, and that of nitrogen 1.25 kilograms per cubic meter. Since air is 4/5 nitrogen and 1/5 oxygen, the density of air under standard conditions is 1.29 kilograms per cubic meter.

Low as this density is, and accustomed as we are to thinking of air as something we can ignore, there are substantial masses involved. An American living room of moderate size (12 feet by 18 feet, and with a ceiling 8 feet high) has a volume of 49 cubic meters. The air in such a living room has a mass of 63.4 kilograms, which is the average weight of an American adult. (If the living room were filled with uranium hexafluoride vapor, that vapor would have a mass of 790 kilograms.)

There are several gases that are less dense than air. Ammonia has a density, under standard conditions, of 0.77 kilograms per cubic meter, methane one of 0.72 kilograms per cubic meter, helium one of 0.18 kilograms per cubic meter, and hydrogen one of 0.09 kilograms per cubic meter.

Gaseous hydrogen, then, is just under 1/4 as dense as air, and is 1/250,000 as dense as osmium.

So far, though, we have dealt only with substances here on the surface of Earth. Suppose we spread our sights broader than that. In so doing we will eventually encounter densities both higher and lower than anything on Earth's surface.

## PRESSURES ON EARTH

Under the influence of a gravitational field, everything is pulled in a direction which seems "down," and this gives rise to a sensation of weight. This weight is distributed over an area, and the quantity of weight per unit area is "pressure."

We ourselves live at the bottom of an ocean of air—the atmosphere—which is pulled downward by Earth's gravitational field. Its weight presses upon us in all directions, subjecting us to "air pressure." We are not aware of this air pressure under ordinary circumstances, because the fluids within our tissues push outward with a pressure just equal to that of air pressure, neutralizing the effect. Through most of history, therefore, human beings have been unaware of the very existence of the phenomenon.

Still, it was noticed at various times in the past that no matter how well-built a pump, and no matter how assiduously people worked the pump handle, water could never be raised more than a little more than 10 meters above its natural level.

Various suggestions were offered to account for this, and in 1643, an Italian physicist, Evangelista Torricelli (1608–47), looked into the matter. Suppose, he thought, air, for all it seemed so nearly nothingness, actually had weight, even as other things had. If so, it would exert a pressure, and it would be this air pressure that pushed the water above its natural level. Perhaps there was only so much air, and therefore only so much air pressure, and the total air pressure was only sufficient to balance a column of water a little over 10 meters high.

To check this thought, Torricelli made use of mercury, the density of which is nearly 13.5 times that of water. If the air pressure could balance a column of water a little more than 10 meters high, it should also suffice to balance a column of mercury about 3/4 meter high. That column of mercury would weigh as much as the much taller column of the much less dense water.

Torricelli took a 1.3-meter-long piece of glass tubing, closed at one end. He filled it with mercury, corked it, upended it into a large dish of mercury, and uncorked it. The mercury began to empty out of the tube as one might expect, but it did not do so altogether. A column of mercury 0.76 meters high remained in the tube, supported by the weight of the air pressing down on the mercury in the dish.

Thus, the pressure of the atmosphere was equal to that of a column of mercury 0.76 meters high.

A column of mercury which was 1 square meter in cross-sectional area and 0.76 meters high would weigh 10,332 kilograms, something that is easy to determine. This means that the weight of all the air over 1 square meter of Earth's surface at sea level would also weigh 10,332 kilograms.

We can say, then, that air pressure is equal to 10,332 kilograms per square meter, and we can refer to 10,332 kilograms per square meter as "1 atmosphere."

We are talking here, mind you, of a kilogram of weight, which is not the kilogram used in the metric system, where it is a unit of mass. Weight is a force—the force with which something presses against something else under the pull of a gravitational field.

The fundamental property of a force is that of being able to accelerate a mass.

Suppose we begin with a kilogram of mass at rest in a vacuum. Imagine a force capable of setting that kilogram into motion and, as it continues to be applied, of making it move faster and faster. The force might make the kilogram move at a velocity of 1 meter per second at the end of one second, 2 meters a second at the end of a second second, 3 meters a second at the end of a third second, and so on.

The size of that force is then said to be 1 kilogram-meter per second per second. To avoid having to use that eleven-syllable phrase perpetually, scientists define a force of 1 kilogram-meter per second per second as equal to "1 newton," in honor of Isaac Newton, who first worked out this connection between forces and accelerations.

A kilogram of weight is equivalent, in force, to just about 9.806 newtons. Therefore, air pressure at sea level is equal to 101,320 newtons per square meter.

Scientists have defined 1 newton per square meter as "1 pascal," in honor of the French physicist Blaise Pascal (1623–62), who conducted important experiments on the matter of air pressure after Torricelli's fundamental discovery. Consequently, air pressure at sea level, or 1 atmosphere, is equal to 101,320 pascals.

Air pressure is not the greatest pressure which a human being can experience. A column of water 0.332 meters high exerts 1 atmosphere of pressure. If, then, someone dives into a lake to a depth of 10.332 meters, the pressure on his body is a total of 2 atmospheres.

The body can endure that because the internal pressure of tissue fluids rises to match the additional external pressure. This, in theory, can go on indefinitely as one dives deeper and deeper, but complications ensue, as we saw earlier in the book, that limit the process.

The extreme depth of the ocean, in the Marianas Trench, is a little over 11 kilometers. Ocean water is salty and is, therefore, denser than fresh water, and so a column of a given height is attracted more strongly by gravity and exerts more pressure than a similar column of fresh water does. Add to that the fact that, with depth, water is slightly compressed and the density goes up a tiny bit further. Taking all this into account, the pressure in the deepest part of the ocean is about 1,070 atmospheres or 108,000,000 pascals.

It is against such pressure that human beings have had to be protected by the strong metal walls of the bathyscaphe, in penetrating to the very bottom of the ocean.

Nor is the pressure at the very bottom of the ocean the greatest that exists on the planet Earth.

The solid material that makes up the ball of Earth is denser than water and extends deeper, so that if one imagines oneself sinking deeper and deeper into Earth, one would quickly encounter pressures greater than those anywhere in the ocean.

One thousand kilometers below Earth's surface, the weight of the overlying rock presses down with a pressure of over 40,000,000,000 pascals, or about 400,000 atmospheres, which is 400 times the greatest pressure in the ocean.

Pressure continues to climb as we probe deeper into Earth. The substance of Earth is rock of one sort or another until a depth of 2,900 kilometers is reached, and then the rock gives way to molten metal.

At the very center of Earth, it is calculated that the pressure is 364,000,000,000 pascals, or about 3,600,000 atmospheres.

Under the compression induced by such pressures, the densities of materials become significantly greater than they would be at sea level, where only 1 atmosphere of pressure is pushing down upon them. The vast pressures deep inside Earth compress the atoms themselves, forcing the electron shells in the outer region to move closer to the nucleus, against the strong electromagnetic forces that tend to keep them spread out.

Rocks with densities of 3,000 kilograms per cubic meter at the surface are nearly 6,000 kilograms per cubic meter at a depth of 2,900 kilometers, where the region of molten metal takes over.

The metal is thought to be chiefly iron with a 10 percent admixture of the related metal nickel. At sea level, such a mixture would have a density of 8,000 kilograms per cubic meter. At a depth of 2,900 kilometers, however, under the pressure of the overlying layers of rock, the metal has a density of about 9,700 kilograms per cubic meter. This value rises still further at greater depths until, at Earth's very center, where pressure is the highest, the density of the metal core is thought to be about 13,000 kilograms per cubic meter.

And yet even at Earth's center, the density of the material present there is only half that of a nugget of pure osmium at Earth's surface. Ordinary osmium retains the record density for Earth in all its parts, and not on the surface only. (Of course, if a quantity of osmium were present in bulk at Earth's center—which it is not—its density might reach a mark of 35,000 kilograms per cubic meter under the pressures existing there.)

Human beings are not likely, in the foreseeable future, to penetrate the depths of Earth and to experience the huge pressures, and their effects, in person. Might it be possible, however, to bring high pressures into the laboratory?

The French physicist Emile Hilaire Amagat (1841–1915) was a pioneer in this respect. By applying mechanical pressure to a small volume, and by devising seals that were particularly efficient, he managed to reach pressures as high as 3,000 atmospheres in the 1880s. What stopped him at that point was that even the best seals he could devise eventually gave way—but it was no mean accomplishment. Amagat reached pressures three times those in the deepest portion of the oceanic abyss.

In 1905, the American physicist Percy William Bridgman (1882–1961) was working for his Ph.D. at Harvard and was studying the behavior of certain optical phenomena under the influence of pressure. He began to interest himself in the problem of reaching higher and higher levels of pressure and worked out ingenious seals that would retain fluid under more and more extreme conditions.

He soon reached a pressure of 12,000 atmospheres and then, in successively improved devices, went to 20,000, then 30,000, then 50,-000, then 100,000, and then, finally, to an occasional 425,000 atmospheres, which is almost 1/8 of the pressure at Earth's center.

In recent years, Peter M. Bell of Carnegie Institution has made use of a device that squeezes material between two diamonds (the hardest substance known) and, in this way, has managed to reach pressures of 1,500,000 atmospheres, over 2/5 that at Earth's center.

Even Bell's figures have been outdone, at least temporarily, at the California Institute of Technology, where projectiles were fired at high velocities from a cannon. At impact point, momentary pressures of several million atmospheres—approaching that at the center of Earth—can be attained.

In this way, it is possible for scientists to study the changes that go on under high pressure in the metals and minerals that make up the major portion of Earth and thus get a better idea of the structure of the interior of Earth—and of other planets.


## PRESSURES BEYOND EARTH

Even if we do surpass the pressure at Earth's center, thus setting a planetary record, it is by no means a record for the solar system. There are four planets more massive than Earth, and each of these must, of necessity, have a greater central pressure than Earth does, since, in each case, there is a greater mass of material being compressed by a more intense gravitational field.

The most massive of the planets is Jupiter, and some estimates place its central pressure as high as 100,000,000 atmospheres, or thirty times that at the center of Earth.

At the pressures existing at Jupiter's center, the atoms themselves are undergoing a strain that threatens their ability to resist gravitational compression. With the development of the concept of the nuclear atom by Rutherford in 1911, it became clear that atoms might, under sufficient pressure, break down.

Under ordinary pressures, almost all the mass of an atom is concentrated in the tiny atomic nucleus at the center. These nuclei are surrounded by comparatively large volumes filled with electrons. In these circumstances, the nuclei cannot approach each other; the electrons are in the way.

With sufficiently high pressure, however, the electron structure in the outskirts of the atoms breaks down, and the bare nuclei are exposed. Now nuclei can approach each other, collide, and interact. If most of the material in a massive body is hydrogen, the nuclei are single protons, and these can "fuse" on interaction to form helium nuclei, liberating a quantity of heat. As a result of such fusion at the center of a massive body (one that is substantially more massive than Jupiter), enough heat is produced to allow the entire body to radiate light. In short, a star is formed.

This is why the sun glows with its own incandescent light.

The sun is 1,020 times the mass of Jupiter, and the pressure at its center is estimated to be as high as 330,000,000,000,000 atmospheres, which is about 3,300 times that at the center of Jupiter. This is surely enough pressure to break down the atomic structure at the sun's core and produce nuclear fusion.

Once the atomic structure breaks down to produce what some term "degenerate matter," the massive nuclei, moving closer together than they ever can when atoms are intact, produce a substance with far higher than normal density. The density at the sun's center, for instance, is estimated to be about 160,000 kilograms per cubic meter, or seven times the density of osmium.

The material at the center of the sun is mostly helium, the nuclei of which have an atomic weight of 4 as compared with osmium's 190. For the light helium nuclei to produce a material that is seven times the density of osmium, the helium nuclei at the center of the sun must be pushed to a distance of separation only about 1/7 that of osmium nuclei at Earth's surface.

Since the distance between nuclei in ordinary intact atoms is something like 200,000 times the nuclear diameter, cutting that distance to 1/7 still leaves nuclei separated, on the average, by nearly 30,000 times their own diameter. In proportion, this would be similar to Ping-Pong balls scattered about with an average separation of a little more than a kilometer. Such a separation would allow the Ping-Pong balls to move about freely, and without any appearance of being crowded.

In spite of the enormous density at the sun's center, therefore, the material there acts like a gas.

But then, the sun does not set a record in this respect. There are some stars that are more massive than the sun, even up to fifty times as massive. All of these may be expected to have central pressures and densities greater than those of the sun.

These, too, are not record holders—by far.

All that keeps stars from collapsing under their own enormously intense gravitational fields is the equally enormous central temperatures generated by nuclear fusion. Eventually, though, the nuclei that undergo fusion are consumed and only the large nuclei that are the products of fusion are left. Eventually, after anywhere from millions to hundreds of billions of years of shining, a star can no longer evolve enough heat to remain expanded—and it then collapses.

In the case of these collapsed stars, virtually all the material is made up of degenerate matter, instead of only the core material, as in the case of the sun. Pressures and densities go up much higher.

The first collapsed star to be discovered was a companion star of Sirius (something mentioned earlier in the book). It was discovered in 1844 by Bessel. He noticed that Sirius was moving in a wavy path, as though something with the gravitational intensity of a star was pulling at it. Bessel could not see any star where one should have been and assumed the companion of Sirius to be a dead star, one that had ceased shining.

The companion was finally seen in 1852, by an American astronomer, Alvan Graham Clark (1832-97). It shone faintly and appeared to be, if not a dead star, a dying one. From its movement, and that of Sirius, it could be seen that the companion star ("Sirius B") had a mass equal to 1.05 times that of our sun.

Then, in 1915, the American astronomer Walter Sydney Adams (1876-1956) was able to take the spectrum of Sirius B. From that spectrum, it could be shown that it had a temperature as high as that of Sirius itself. Its surface was hotter than that of our sun.

In that case, though, why did it appear so faint in the sky? If it was as hot as Sirius itself and nearly as massive, why wasn't it nearly as bright as Sirius in appearance?

The only answer possible was that it was much smaller than Sirius in diameter. Its surface might be hot and bright, but there was very little surface. In fact, from present-day measurements it would appear that the diameter of Sirius B is only about 11,100 kilometers. Sirius B is therefore smaller than Earth, and yet it is a white-hot star. Combine its brilliance with its small size and it is no wonder that it is called a "white dwarf."

Yet for all its small size, it still has a mass 1.05 times that of the sun. Clearly, it was once an ordinary large star that ran out of fuel and could not remain expanded against the pull of its own gravity. There are a considerable number of white dwarfs in existence, all of them the collapsed remnants of stars.

If you consider an object that is 1.05 times the mass of the sun compressed into a ball smaller than Earth, you can see that the overall density of Sirius B must be enormous. In fact, its average density is 2,900,000,000 kilograms per cubic meter, and it is estimated that at its center, the density may be as high as 33,000,000,000 kilograms per cubic meter. Such a density is roughly 1,000,000 times that of osmium.

Even in the case of such an enormous density, the nuclei at the core of a white dwarf are separated from each other by a distance equivalent to Ping-Pong balls with an average distance of 15 meters between them. The material would *still* behave as though it were a gas.

The more massive a star originally, the more intense the gravitational field, and the more catastrophic the eventual collapse. If the star is massive enough, the collapse can proceed to the point where the nuclei approach each other and make contact.

Such stars are "neutron stars," and these were first detected in 1967 by a young British student named Jocelyn Bell, who was working for the astronomer Antony Hewish (1924–    ).

A neutron star with the mass of the sun would have a diameter of no more than 14 kilometers. Its density would be the density of the atomic nucleus itself—in fact, a neutron star could almost be looked upon as a giant atomic nucleus.

The density of a neutron star would be something like $10^{18}$ kilograms per cubic meter, or at least 100,000,000 times that at the center of a white dwarf, and 50 trillion times that of osmium. The material in a neutron star would not act like a gas, but, despite an enormous temperature, would behave like a solid.

Yet even a neutron star is not the ultimate. If a star is massive enough and if it collapses forcefully enough, it may smash through the nuclear barrier, too. Even the substance of the atomic nucleus would not be able to withstand gravitational compression. The nuclear particles themselves would break down and there would be nothing left that could keep the process of collapse from continuing indefinitely.

The result is a "black hole," in which matter simply collapses and collapses, with pressures and densities and gravitational intensities rising, without limit, toward the infinite.

Most astronomers feel fairly certain that black holes exist and that some may even have been detected.


## TOWARD VACUUM

Suppose we move in the other direction.

The least dense substance at sea level is the gas hydrogen, which has a density of 0.09 kilograms per cubic meter. This is equivalent to saying that a cubic meter of hydrogen contains $2.7 \times 10^{25}$ (27 trillion trillion) hydrogen molecules altogether.

This is more or less true of any gas at sea level. Thus, air at sea level contains 27 trillion trillion molecules, too. Air is, of course, a complex mixture of many gases. Ninety-nine out of every hundred molecules are either oxygen molecules or nitrogen molecules, but that remaining, one out of each hundred, is still enough to represent large numbers of atoms of even the rarest substances. For instance, only $6 \times 10^{-19}$ (less than a quintillionth) of the particles of the atmosphere are atoms of the rare gas radon. Nevertheless, this means that a cubic meter of air at sea level contains 16,000,000 atoms of radon.

Even though there are so many molecules of air in a cubic meter, the individual molecules are so small that the average distance between them is just about 100 times the diameter of the individual molecules. (This is like dealing with Ping-Pong balls spaced out with an average separation of 1.5 meters.)

The separate molecules in gases are not held together by strong chemical bonds, such as are to be found in solids. They are held together by virtually no internal forces at all—only by the pressure of the gases higher up, gases which are pulled down by Earth's gravity. If air pressure were somehow reduced, the molecules of air would spread apart and grow less dense. If air pressure were reduced sufficiently, the molecules of air (or any gas, for that matter) would spread apart indefinitely and vanish into near nothingness.

One way of reducing the air pressure is to imagine the gravitational intensity decreased. If it decreased sufficiently, it would fail to produce sufficient pressure to keep the atmosphere from spreading apart indefinitely and leaking away into outer space. That is why worlds such as the moon and Mercury have no significant atmosphere. Nor can anything still smaller than the moon have one.

Another way to reduce air pressure is to imagine ourselves moving up through the atmosphere to greater and greater heights. In that case, we are leaving more and more of the atmosphere below us and the air pressure we would experience would depend only on the fraction of the atmosphere still above us. This means that air pressure should drop steadily as we move upward.

As mentioned earlier in the book, this was first demonstrated in 1646 by Pascal.

If the atmosphere were equally dense all the way up, then it would come to an end at 8 kilometers, for it takes only a column of air 8 kilometers high, all of it being at sea-level density, to produce the air pressure we actually observe at sea level. However, the higher we go, the lower the air pressure, and the more the molecules of the atmosphere spread apart. The atmosphere gets less and less dense, taking up more and more room, therefore, and stretches far higher than 8 kilometers, though at the price of getting steadily more rarefied.

At the top of Mt. Everest (a height of 8.84 kilometers) the air pressure is only 0.31 what it is at sea level—or about 31,400 pascals. (This is something which adds immensely to the difficulty of climbing that final mile when high peaks are scaled.)

Scientists have made use of balloons and rockets to determine the properties of the atmosphere at heights far greater than any mountains. At a height of 10 kilometers, the air pressure has dropped to 28,000 pascals; at a height of 50 kilometers to a mere thousandth of the pressure at sea level, or 101 pascals, at 100 kilometers to 0.08 pascals, and at 220 kilometers to 0.00002 pascals.

As far as any practical use of the atmosphere is concerned, we might suppose that by the time we reach a height of 50 kilometers, what wisps of air are left are practically nothing, and might as well be considered a vacuum.

Nevertheless, even at 220 kilometers, where the faint wisps of air have less than five-billionths the density of air at sea level, there are still $5 \times 10^{15}$, or 5 quadrillion, atoms or molecules in a cubic meter. That is enough to interfere significantly with the flight of artificial satellites and, through the resistance they impose, to rob them, eventually, of so much energy as to bring them to the ground—as in the case of the ill-fated *Skylab*.

As one moves farther and farther from Earth, the gas density continues to drop, but it never reaches zero! The sun is forever giving off quantities of subatomic particles, and these spray out at high speeds in all directions ("the solar wind"), and so the regions of space surrounding the Earth-moon system contain 5,000,000 to 80,000,000 particles per cubic meter (mostly protons—the nuclei of hydrogen atoms).

This still sounds like a great deal in numbers, but this is less than a millionth of a trillionth of the density of air at sea level.

What about interstellar space, the space between the stars? There, the distribution of matter is not even. There are dust clouds, for instance, that may fill 1/25 of the total volume of space in the spiral arms of the galaxy, and these may consist chiefly of protons, if hot stars are within them or near them, or of intact hydrogen atoms if there are no nearby stars to supply the energy to disrupt them.

The total particle content in such clouds may be up to 100,000,000 per cubic meter, and since the 1960s, the radio waves they emit have shown that there are small numbers of molecules, including some complex carbon compounds, which are also present.

In the regions between the clouds, where density is at a minimum, there may be as few as 100,000 hydrogen atoms per cubic meter. Even here, though, the various "stellar winds" contribute. In intergalactic space, far from all stars, there may be as few as 0.1 hydrogen atoms per cubic meter.

Nevertheless, if you took a volume of the very emptiest intergalactic space, a volume that was the size of the planet Earth, it would contain a total of $10^{19}$ (10 million trillion) hydrogen atoms. Space is *never* truly empty.

Human beings have learned to create vacuums here on Earth, too. The first of significance was produced by Torricelli, when he up-ended his column of mercury and let part of it run out. The space it vacated at the closed top of the tube contained nothing but traces of mercury vapor, and this is a "Torricellian vacuum."

The mercury vapor keeps it from being a perfect vacuum, and, at ordinary temperatures, a cubic meter of Torricellian vacuum would contain $3.5 \times 10^{19}$ mercury atoms per cubic meter.

Since Torricelli's time, many ways of pumping air out of a closed container have been developed, and vacuums containing less than ten-trillionths the number of particles in a Torricellian vacuum have been produced.

A cubic meter of even the best man-made vacuum, however, still contains about 3,000,000 atoms or molecules per cubic meter. This is on a par with interplanetary space, but is far from being as good as the emptiest interstellar space (let alone the emptiest intergalactic space).

# PART IV
# The
# Horizons
# of
# Energy

# 21
# High Temperatures

## MELTING AND BOILING POINTS

Just as density is a measure of the concentration of mass in a given volume, so is temperature a measure of the amount of energy in a given volume. Temperature, the measure of heat intensity, is a strong influence on events, and is something of which everyone is continually aware.

Human knowledge of temperature exists as a biological sensitivity. Objects are cool or warm to the touch, depending on whether heat flows out of the body into an object, or out of the object into the body. In the first case, body temperature drops at the point of contact; in the latter case, it rises. In either case, the brain interprets the event appropriately.

Heat or cold can also be felt at a distance, as when standing before a roaring fire, or before a mount of ice. In the first case, heat is carried from the fire to the body by air currents and radiation, while in the latter case, heat is carried from the body to the ice.

Attempts to measure temperature objectively date from 1593, when Galileo invented the first "thermometer" (from Greek words for "heat measure"). By that time it was realized that objects tend to expand with rising temperature, and that the extent of this expansion might be used as a measure of the temperature.

Galileo warmed a hollow bulb, to which a long hollow stem, open to the air, was attached. The bulb was inverted, and the stem was dipped into water. The bulb cooled and, as it did, the air within contracted, and water moved up the stem. Thereafter, as temperature went up and the air in the bulb expanded, the water level dropped; as temperature went down, the water level rose. It was a crude affair, since air pressure outside, as it varied, also sent the water level up and down, but it was a beginning.

In 1654, Grand Duke Ferdinand II of Tuscany (1610–70) was the first to use a bulb to which a sealed stem was attached. The bulb was filled with alcohol, and its rise and fall in the stem depended on temperature only and was not affected by air pressure.

The first person to construct a thermometer that was sufficiently accurate to be useful in scientific experiments was the German-Dutch physicist Gabriel Daniel Fahrenheit (1686–1736). In 1714, he used clean mercury to fill the thermometer bulb, sealing it into a vacuum.

He set up a scale against which to measure temperature. He placed the thermometer in a mixture of water, ice, and salt and marked the level of the mercury as 0°. The body temperature of a person in good health, as measured by placing the bulb of the thermometer in the mouth, he placed at 96°. Against this scale it turned out that the freezing point of water is 32° and the boiling point of water is 212°. (The normal body temperature is now taken to be 98.6° on this scale.)

This is the "Fahrenheit scale," and temperatures using it are written as 32° F. or 98.6° F. It is used in the United States, but scarcely anywhere else in the world. Even in the United States, it is now beginning to fade out.

The Swedish astronomer Anders Celsius (1701–44) introduced a scale in 1743 in which the freezing point of water was set at 0° and the boiling point of water at 100°. This was originally called the "Centigrade scale" from Latin words meaning "hundred steps." In 1948, however, it was officially termed the "Celsius scale" in honor of the inventor. Either way, we say 0° C. and 100° C. The Celsius scale is used the world over, increasingly even in the United States, and is used in this book.

It was only after accurate thermometers became available that questions could be asked as to how high a temperature might be encountered.

The normal temperature of our general surroundings fluctuates through the day and night and over the seasons. That temperature is, sometimes, higher than the normal body temperature of 37° C. In New York City, for instance, the highest temperature recorded in Weather Bureau history (which dates back only to the 1890s) was 41.7° C., recorded in July 1966.

On July 10, 1943, the highest temperature in the United States was recorded in Death Valley in California. It was 56.5° C., just about 20 degrees higher than normal body temperature. Even this is not a world record, however, for on September 13, 1922, at Al 'Aziziyah in Libya, the temperature rose to 58° C.

There are, however, temperatures on or near the surface of Earth that are much higher than the general atmospheric temperature about us. Lightning bolts involve enormous temperatures of a very localized and temporary kind. Volcanic eruptions do not produce temperatures nearly as high, but those that are produced are maintained for long periods of time. Either way, forest fires can be produced.

Forest fires have existed as long as forests have, so that fire long antedates humanity. Hominids were, however, the first organisms who made deliberate use of fire—who did anything other than flee from it. As long as 500,000 years ago, *Homo erectus,* a relatively small-brained ancestor of modern humanity, seem to have been sitting around campfires.

*Homo erectus* (as well as *Homo sapiens* for many thousands of years) could only use fire which had been rescued from material burning from natural causes, generally a lightning stroke, and had to keep it painstakingly alive by feeding it fuel. If the fire went out, a new light had to be gathered from some other campfire or, if none existed within reach, a new lightning-generated fire had to be awaited.

It was not, perhaps, till 7000 B.C. that human beings learned techniques for starting fire where none had existed before, usually through friction.

The ordinary fire used by early man is the result of a chemical reaction in which vapors, arising out of heated organic materials such as wood or oil, combine with the oxygen in the air. This combination produces enough concentration of heat to radiate light, and the temperatures that result are in excess of 1,000° C.

By using fire, human beings could therefore produce temperatures that would, in turn, induce effects that would not take place without it. For instance, a kettle of water placed over a fire will surely boil, though no purely meteorological factor will cause it to do so. Through the heat of fire, food can be cooked, sand turned into glass, clay into brick, various ores into metals.

An ordinary wood fire is by no means as hot as a fire can be. In modern times, much more energetic chemical reactions have been taken advantage of. The combination of the gases oxygen and hydrogen yields a flame with a temperature of 2,800° C.; that of oxygen and acetylene, 3,300° C. Far higher temperatures can also be attained (briefly) these days.

The combination of high temperatures and accurate methods for measuring them (involving devices other than ordinary thermometers) yielded accurate figures for high melting points.

Under natural conditions, the only common substance that melts at ordinary temperatures is ice. Early in historic times, however, human beings learned to use fire to prepare metals for ore and, in the process, metals were melted.

Of the seven metals known to the ancients, one was mercury, which is already liquid at all ordinary temperatures. Lead and tin, while solid at ordinary temperatures, could be melted without undue difficulty by even quite primitive methods. The melting point of tin is 231.9° C. and that of lead is 327.5° C.

Silver, gold, and copper have higher melting points: 960.8° C., 1,063° C., and 1,083° C., respectively. The highest melting point among all the ancient metals is iron, which does not melt until a temperature of 1,535° C. is reached.

The melting points given above are for the pure metals. Metals that are mixed with each other (alloys) tend to melt at lower temperatures than pure metals do, but even so, iron alloys remain more resistant to melting than other alloys do. It is partly for this reason that iron came to be used so much later than the other metals, despite the fact that iron is by far the most common of them, as well as the most useful.

Thus, bronze (an alloy of copper and tin) came into use in some places as early as 3600 B.C., and was the first metal hard enough to serve for tools and weapons. Iron, on the other hand, was not successfully smelted until about 1400 B.C., more than 2,000 years later. Higher temperatures were needed than wood fires could supply, and the use of charcoal had to be introduced as a fuel.

Rocks are harder to melt than the long-known metals are. Such common components of rocks as aluminum silicate, calcium silicate, and magnesium silicate melt in the neighborhood of 2,000° C.

In modern times, however, metals have been discovered that melt at higher temperatures still. Hafnium melts at 2,150° C., ruthenium at 2,250° C., iridium at 2,410° C., niobium at 2,468° C., molybdenum at 2,610° C., tantalum at 2,996° C., osmium at 3,000° C., rhenium at 3,180° C., and tungsten at 3,410° C.

Tungsten has the highest melting point of all metals, and that is one of the reasons for its use as a filament in incandescent light bulbs. It resists the continuous high temperatures best (provided it is surrounded by a gas, such as argon, with which it does not react).

There are two elements that are not metals but that are also high-melting. Boron melts at 2,300° C., and carbon sublimes (that is, the solid turns directly into a gas, rather than melting into a liquid) at a temperature somewhat higher than 3,500° C. Of all the elements, then, carbon remains solid at the highest temperature.

Atoms of different elements can combine chemically to form high-melting compounds. (These are not the same as alloys, which are simple mixtures, rather than chemical compounds.) The highest-melting compounds are generally made up of a metallic atom in chemical combination with a small nonmetallic atom such as boron, carbon, oxygen, or nitrogen. Boron carbide has a melting point of 2,350° C., niobium nitride one of 2,573° C., and calcium oxide one of 2,580° C.

There are at least nine compounds of this sort with melting points above 3,000° C., and four that equal or surpass 3,500° C. Niobium carbide melts at 3,500° C., zirconium carbide at 3,540° C., tantalum carbide at 3,880° C., and hafnium carbide at 3,890° C.

Of all the substances in existence, then, hafnium carbide seems to be the only one that remains solid at a temperature of 3,890° C. Above that temperature, only liquids and gases can exist.

If the temperature continues to go up, of course, liquids boil and become gases. Platinum, for instance, boils at 4,300° C., and that is by no means the maximum. Osmium boils at something above 5,300° C., while rhenium and tungsten boil at about 5,900° C.

The metal tantalum and the compound tungsten carbide both boil at about 6,000° C., and that is the probable maximum for any substance under conditions like those on Earth's surface. Above 6,000° C., all substances are gases.


## PLANETS AND STARS

What if we move away from Earth's surface? What of the surfaces of other worlds?

In our solar system, there are three large objects that are as close to the sun as Earth or closer than Earth is. These are our satellite, the moon, and the two inner planets, Mercury and Venus.

The moon circles Earth at a distance much smaller (only 1/390) than the distance of either from the sun. This means that the moon is essentially at the same distance from the sun as Earth is and might be expected to experience the same surface temperature, by and large, as Earth does.

However, Earth has an atmosphere which conserves and circulates heat, so that the day is not as hot, or the night as cold, as they would be without an atmosphere. The moon does not have an atmosphere and therefore experiences greater extremes of surface temperature. This is all the more true since the moon rotates, with respect to the sun, only once in 29.5 days. It remains exposed to solar heat, in each of its rotations, for a considerably longer interval than Earth does.

The result is that the moon, at its equator, reaches a surface temperature of just over 100° C., a temperature more than 40° C. higher than is reached by any spot on Earth's surface that is heated by the sun only.

Mercury is the closest of the planets to the sun; in the course of its revolution it is at its nearest only 46,000,000 kilometers from the sun, a distance only 0.3 times that of Earth from the sun. What's more, it has no atmosphere and rotates very slowly. The maximum surface temperature of Mercury reaches about 425° C. If there were any metallic tin and lead on Mercury's surface (which there isn't), they would melt in the full heat of the noonday sun at Mercury's closest approach.

Nevertheless, Mercury does not hold the record for surface temperature among the planets of the solar system.

Venus, at a distance of 108,000,000 kilometers from the sun, is 2.3 times as far from the Sun as Mercury is (though only three-fourths Earth's distance). What's more, Venus has an immensely dense atmosphere (ninety times the density of Earth's), which keeps the temperature well distributed. It also has an unbroken cloud layer that reflects about three-fourths of the sun's radiation and prevents it from reaching the Venerian surface. For all these reasons, it might seem that Venus's surface temperature might be comparatively mild—certainly less ferociously hot than Mercury's.

Not so! Venus's dense atmosphere is over 90 percent carbon dioxide, a substance that traps solar heat and allows the temperature to build up to high levels. As a result, Venus's surface temperature, all over the planet, both day and night, is about 475° C., 50° C. higher than Mercury at its worst. What's more, while specific spots on Mercury's surface cool off a great deal during the night, no spot on Venus's surface ever cools significantly, thanks to the temperature-equalizing effect of the thick atmosphere.

There is one small body that should experience a higher surface temperature than even Venus does—though only intermittently. That is an asteroid named Icarus, a chunk of rock about a kilometer across that was discovered by the German-American astronomer Walter Baade (1893–1960) in 1948. It revolves about the sun in 1.12 years and at its closest approach to the sun is only 28,500,000 kilometers away, which is only half the distance of Mercury's closest approach.

Undoubtedly, Icarus's surface, exposed to the sun so closely, must for a while at least, be heated to a temperature of as much as 650° C. The surface of Icarus would then be hot enough to give off radiation energetic enough to be detected as light (if one could shield out the unbearable glare of the swollen sun). Icarus's surface would, in other words, be warmed to a dull red-heat.

There are occasional comets which approach the sun even more closely than Icarus does, but whereas Icarus is a lump of rock, comets seem to be icy materials with an admixture of dust and gravel (and, sometimes, with a rocky core). As comets approach the sun, some of the icy material evaporates and the dust and gravel is released. This forms a haze around the nucleus of the comet that keeps that nucleus from reaching the temperature it might otherwise reach. The temperature of a comet at a close approach to the sun is therefore uncertain.

There must also be occasional meteors that approach the sun closely enough to vaporize.

This brings us to the hottest object in the solar system, the sun itself. The surface temperature of the sun can be determined from the nature of its radiation, according to a rule first worked out in 1879 by the Austrian physicist Josef Stefan (1835-93).

The surface of the sun is, for the most part, at a temperature of about 5,500° C. This is just below the boiling point of rhenium, tungsten, tantalum, and tungsten carbide. These substances, however, are undoubtedly present only in traces, if at all, in the sun, and are not present in sufficient quantity in any one place to appear as liquid. There is no question, then, that the sun's surface is entirely gaseous in nature.

This is true even though there are places on the sun's surface where the gases expand and cool through some local action (as yet not clearly understood) that involves the sun's magnetic field. These cooled gases do not radiate as strongly as do the surface areas generally, so they seem dark against a brighter background. The temperature at the center of these "sunspots" can be as low as about 3,750° C. Under such conditions, tantalum carbide and hafnium carbide (if there were a sufficient quantity in one spot, which there is surely not) would actually remain solid.

There are the reverse of sunspots. In the vicinity of sunspots, there are apt to burst out "solar flares," sudden explosions of energy that last anywhere from a few seconds to nearly an hour. These gleam whitely against the already-bright background of the sun because they are much hotter than the solar surface generally.

And what about other stars? The vast majority of the stars we can see in the sky, either by our unaided eye or by our instruments, are, like the sun, "main-sequence stars." These are stars in their vigorous youth, shining steadily for long periods of time, thanks to the presence of ample quantities of hydrogen, which serves as raw material for energy-yielding nuclear fusion.

Stars more massive than the sun are hotter than it is. The most massive of the stars on the main sequence, and therefore the hottest, may have surface temperatures of up to 40,000° C.

## INTERIORS

So far, we have talked only about the temperatures at the surface of astronomical bodies. The surface temperature of any object, however, is not likely to be its hottest portion. In many cases, it is its coolest portion.

Consider Earth, for instance. As one moves upward through the atmosphere, away from the planetary surface, the density of the atmosphere drops. So does the total heat content of the atmosphere.

However, the drop in density and the drop in total heat do not necessarily match each other. There are regions where the density drops significantly faster than the heat content does, so that although the total heat is less, the individual atoms and molecules possess more for their individual shares.

Up to a height of 150 kilometers above the surface, the temperature of the atmosphere drops, but above that height, the density drops, and continues dropping, in such a way that individual atoms and molecules get steadily increasing shares of the total heat (which is also diminishing, but not as rapidly).

By the time a height of 300 kilometers is reached, the temperature of the thin wisps of air present is about 1,500° C. (For that matter, a lightning stroke can produce a momentary temperature of up to 30,000° C.)

That sounds as though the upper atmosphere is hot enough to melt iron, but it isn't really. The temperature, meaning the heat intensity per individual atom, is high, but there are so few atoms that the total heat content is very low. A rocket or spaceship passing through the upper atmosphere meets very energetic atoms, but so few of them that the heat transferred from them to the ship is not sufficient to do it any harm.

The same thing happens in the case of the sun, but to a more extreme extent, naturally. The upper atmosphere of the sun is its "corona," usually visible only during a total eclipse. The temperature of the corona mounts to as much as 1,000,000° C., something that was discovered in 1942 by the Swedish astronomer Bengt Edlen (1906–      ), from the nature of the radiation emitted by the corona.

Presumably, the coronas of the hottest stars would be hotter in proportion and might reach a temperature of 10,000,000° C.

Not only does moving up from the surface bring one to regions of higher temperature, moving down from the surface does the same. Consider, once more, Earth.

The solar system formed from a cloud of gas and dust that condensed under the influence of its own gravitational field. Most of the original cloud condensed into what is now the sun, but some of the material at the outskirts coalesced into fragments of matter that gathered together to form the planets and satellites that now exist.

The gathering matter possessed kinetic energy, and as the matter collided and came to rest (relative to each other) that kinetic energy was converted into heat. Earth and other bodies of considerable size began their existence as quite hot objects, therefore. The outer surface cooled off, but the interior remained hot.

The interior remained hot over a period of billions of years, because, first, the outer, cool layers of rock serve as an excellent heat insulator, cutting down the flow of heat from the hot interior to outer space to a comparatively small trickle; and, second, the presence of radioactive atoms of uranium, thorium, and of certain varieties of potassium and samarium within Earth's rocky structure produced heat as they broke down—not much, but enough to balance the small quantities of heat that were lost. Therefore, Earth has remained hot at the center through all its 4,600,000,000-year history, and is hot today.

We know from our experience with mines that the temperature seems to rise about 1 degree for every 30 meters we dig downward. This rate probably slows up below the crust and upper mantle, where most of the radioactive elements are to be found. At the center of Earth, according to reasonable estimates by geologists, the temperature may be 4,000° C.

If this is so, the center of Earth is as hot as the cooler portions of the solar surface.

Not only is the center of Earth hotter than the uppermost wisps of atmosphere (though not as hot as a lightning flash), but while the upper atmosphere is almost a vacuum, Earth's center is densely packed with matter. The total heat in Earth's center is therefore many millions of times as great as that of the upper atmosphere.

Naturally, other astronomical bodies are hot at the center, too. In general, the more massive the body, the hotter it is at the center, since more kinetic energy has been stored as heat in the course of the process of formation. The only planetary bodies we would expect to be hotter at the center than Earth is are those that are more massive than Earth, and only four of these are known: Jupiter, Saturn, Uranus, and Neptune.

The largest of the planets, by far, is Jupiter. Whereas it is only 6,378 kilometers from the surface of Earth to its center, it is 71,600 kilometers (eleven times the distance) from the surface of Jupiter to its center.

The visible surface of Jupiter (actually a cloud layer) is cold, far colder than Earth's surface, since Jupiter is five times as far from the sun as we are and receives only 1/25 the radiation we do. As we picture ourselves sinking below Jupiter's visible surface, however, the temperature rises rapidly.

The Jupiter probes of the 1970s have given us the kind of data we need to make reasonable estimates of the internal temperature of that giant planet. At a position 1,000 kilometers below Jupiter's visible surface, the temperature already approaches that at Earth's center, and at greater depths it goes higher still. At the very center of Jupiter, the temperature is 54,000° C., nearly ten times as high as at the surface of the sun. This is still not high enough, however, to initiate nuclear fusion, and so Jupiter remains a planet. It is not massive enough to be even the most dwarfish of dwarf stars.

The sun is 1,040 times as massive as Jupiter. As the sun formed, the temperatures at its center rose considerably higher than they did in the case of the much smaller Jupiter. The temperature rose high enough, indeed, to initiate nuclear fusion. The heat developed by fusion raised the temperature higher still.

In the early 1920s, the English astronomer Arthur Stanley Eddington (1882–1944) was able to show, from an analysis of the amount of heat required to keep the sun from collapsing under the pull of its own gravity, that the central temperature of the sun must be 15,000,000° C.

This is the hottest matter-dense region in the solar system. It has recently been found that there are regions in the neighborhood of Jupiter and of Saturn that are hotter still, thanks, in all probability, to the action of those planets' enormous magnetic fields. There, however, the total number of atoms (energetic though they be) is so small that the total heat is not significant. Moreover, the atoms are too far apart to strike each other often enough to maintain a fusion reaction, even though they are, in theory, energetic enough to do so.

Oddly enough, human beings can now produce temperatures higher still.

Ever since the 1950s, there have been efforts to produce controlled fusion as a source of energy here on Earth. For that purpose, small quantities of hydrogen must be heated to very high temperatures, temperatures higher than those at the center of the sun.

Whereas a temperature of 15,000,000° C. is sufficient to keep fusion going at the sun's center where the material is not only very hot, but very dense, and where the density is maintained by the sun's mighty gravitation, things are different on Earth. Here on Earth, we can't maintain hydrogen density through an intense gravitational field, and if hydrogen must be dealt with in a less dense state, the temperature must be raised still further to compensate.

Human beings can produce extremely energetic atoms (which is equivalent to producing very high temperatures) by accelerating them in various devices that make use of electromagnetic fields, or by heating them very suddenly with powerful beams of laser light. In this way, temperatures of more than 50,000,000° C. have been produced in very small quantities of hydrogen for very small periods of time. At the moment of writing, such temperatures are still not high enough to induce controlled fusion.

Uncontrolled fusion in the form of hydrogen bombs initiated by a triggering uranium fission bomb has been a reality for thirty years now, however. At the center of a hydrogen bomb, it is estimated that temperatures of up to 400,000,000° C. are briefly attained.

Thus, man-made temperatures far outstrip anything nature can do in the solar system, and yet what we do is no match for what nature can do *beyond* the solar system.

The more massive a star, the hotter it is on the surface and at all stages to the core. We can therefore expect a very massive star to have a much higher central temperature than the sun has. Furthermore, all stars tend to grow hotter at the core as they evolve.

The physicist Hong-Yee Chiu, some years ago, tried to calculate the maximum temperature a star's center could have. He suggested that 6,000,000,000° C. would be the hottest temperature we could expect to find in association with stars. That would be the temperature at the core of a star much more massive than the sun, at a time when that massive star had reached the stage in its life cycle where it was ready to explode into a vast "supernova."

And yet even the 6,000,000,000° C. mark does not represent the maximum. Suppose we look back in time toward the birth of the universe in the course of the big bang.

The farther back in time we imagine ourselves, the smaller the volume of the universe and the more tightly packed its energy content (which would remain constant, in total, with time). Each cubic meter of the universe would contain more and more energy, therefore, as we imagined ourselves moving back in time. This is another way of saying that the universe grows hotter and hotter as we go back in time.

Current theories (if they are correct—and this is, as yet, by no means certain) make it seem that we can make sense out of events taking place a very small fraction of a second after the big bang. For instance, at 1/10,000 of a second after the big bang, neutrons and protons seem to have formed out of the still more fundamental particles called "quarks." At that time the temperature must have been 1,000,000,000,000° C. (a trillion degrees).

The closest to the big bang that physicists can push their calculations is $10^{-43}$ seconds, at which time the temperature may have been 100,000,000,000,000,000,000,000,000,000,000° C. (a hundred million trillion trillion degrees).

As we approach still closer to the big bang, the volume of the universe apparently approaches zero, and its temperature rises without limit toward the infinite.

# 22
# Low Temperatures

## PLANETARY SURFACES

Having reached temperatures inconceivably high, let us return to the ordinary temperatures we experience here on Earth's surface and ask how far we can go, and have gone, in the other direction, toward the very cold.

People who live outside the tropics experience cold weather in the winter, and if they live in high enough latitudes or at high enough altitudes, they find that temperatures below the freezing point of water are not uncommon. In that case, the Celsius scale moves into negative numbers. Water freezes at 0° C., after all, and anything colder than that must be negative.

The Fahrenheit scale puts the freezing point of water at 32° F., and so the figure remains positive for 32 Fahrenheit degrees below freezing. Since 32 Fahrenheit degrees are equal to 17 7/9 Celsius degrees, 0° F. is equal to almost −18° C. That is not enough to save us from negative numbers. Winter weather can take us into negative numbers on the Fahrenheit scale, too.

For instance, the coldest New York City day in Weather Bureau history was recorded on February 9, 1934, when the temperature reached −26° C. (the equivalent of −15° F.)—and New York City is not a particularly cold city. On that same day, Rochester, New York, reached a low of −33.3° C.

The lowest temperature recorded anywhere in the forty-eight contiguous states was at Rogers Pass, Montana, on January 20, 1954, when the temperature reached −56.5° C. Naturally, we would expect that Alaska would do better than that, and at Prospect Creek

Camp, Alaska, on January 23, 1971, a temperature of −62° C. was recorded.

Alaska is not the coldest place in the world, either. Siberia is colder, and the towns of Verkhoyansk and Oymyakon in Siberia have both experienced temperatures of −68° C., the latest occasion being in Oymyakon on February 6, 1933. That is a record for the Northern Hemisphere.

In the Southern Hemisphere lies Antarctica, the icebox of the world. The Soviet Union maintains a weather observatory at a spot called Vostok, which is farthest inland and therefore expected to be coldest. On August 24, 1960 (in the Southern Hemisphere, the coldest months are July and August) the temperature reached −88° C., the coldest natural temperature recorded anywhere on Earth.

It would be nice to get rid of negative temperatures, if we could, and, as it happens, we can. In 1699, a French physicist, Guillaume Amontons (1663–1705), discovered that gases contracted at a uniform rate as the temperature dropped. He could, of course, reach only a moderately low temperature, given the state of the art in his time, but he calculated that if the contraction continued at a uniform rate, the gas would reach zero volume somewhere about what we would now call −240° C. His work in this respect was largely disregarded.

In 1802, however, the French chemist Joseph Louis Gay-Lussac (1773–1850) repeated the experiments more accurately and found that gases contracted at a rate that would reduce them to zero volume at −270° C. The present-day value is, as well as can be determined, −273.15° C.

Of course, no one really expects gases to reach zero volume and disappear at −273.15° C. or at any temperature. As the temperature drops, the atoms or molecules gradually approach each other until, eventually, the gas liquefies. At that time, the atoms or molecules are in contact and the liquid contracts only very slightly as it cools further. There is then no "zero volume" to puzzle the mind, and it might seem that the temperature could continue to drop indefinitely.

But then, in 1848, the Scottish physicist William Thomson (1824–1907), who, later in life, received a title and is now better known as Lord Kelvin, showed that the atoms or molecules of all substances, whether gases or not, lost kinetic energy at a constant rate as temperature was lowered. In gases, this showed up as a constant contraction, but that was not the important thing. It was the energy loss that counted. At −273.15° C. the energy content reached zero and the temperature could not be lowered any further. That temperature (equivalent to −459.67° F.) was an "absolute zero."

There is, therefore, an "absolute scale" of temperature in which there are no negative values. In order to turn the Celsius scale into the absolute scale, it is only necessary to add 273 to the Celsius reading (the decimal places are usually ignored).

Thus, body heat, 37° C., becomes 310° K., where "K." stands for Kelvin. Similarly, New York City's lowest recorded temperature of −26° C., becomes 247° K., and the world's record low temperature at Vostok in Antarctica is 185° K.

The surfaces of other worlds of the solar system (excluding the sun and the planet Venus) can reach lower temperatures than Earth's surface ever does.

The moon, which becomes very hot on that part of its surface which is exposed to sunshine, undergoes a rapid temperature drop when that surface moves into the night shadow, since there is no atmosphere to hold the heat. A given spot on the moon's surface remains in night for two weeks, and for that reason, the lunar temperature, just before dawn, has a temperature as low as 100° K., considerably lower than Antarctica's worst.

Mercury, which is far closer to the sun than the moon is, and whose surface can attain far higher temperatures during the Mercurian day, experiences a night that is six times longer than the moon's, and so its temperature drops still lower, despite the nearness of the sun. Just before the gigantic Mercurian sun peeps above the horizon, the temperature of Mercury's surface can be as low as 90° K.

Mars, which is farther from the sun than Earth is, but which has a night that endures just a few minutes longer than Earth's does, does not attain quite such low temperatures as the moon and Mercury. The Martian minimum may be 120° K., which is nevertheless colder than Antarctica.

This is the Martian minimum, attained at its southern polar ice-cap. The top of the cloud layer that serves as the visible surface of Jupiter and Saturn shows temperatures in that same region, not as a record low, but as an average for the entire planet.

The visible surfaces of the planets even farther from the sun have lower temperatures still, as would be expected, since the heating effect of the sun drops as the square of the distance. Uranus is about twenty times as far from the sun as we are, and Neptune is thirty times as far. Uranus therefore gets only 1/400 the warmth from the sun that we do, and Neptune gets 1/900.

It is not surprising, then, that the temperature of Uranus's visible surface is about 90° K., and that of Neptune about 60° K.

Pluto, the farthest of all the known planets, may have a surface temperature of only 40° K. when it is in that part of its orbit that places it farthest from the sun.

## LIQUEFYING GASES

Even at the low temperatures naturally attainable on Earth, certain changes take place that are unusual. For instance, at 233° K. (−40° C.), mercury freezes to a solid.

Again, tin in its ordinary form is called "white tin" and it is then a metal. At low temperatures, however, its atoms rearrange themselves to form a substance called "gray tin" which has nonmetallic properties. The transition point is 286° K. (13° C.), which is only the temperature of a brisk spring day in New York, but at that temperature the transition from white tin to gray tin is extremely slow.

The lower the temperature gets, however, the more rapid the conversion, and at a temperature of 223° K. (−50° C.), white tin crumbles rapidly. This was called "tin disease" by people who wondered what was happening. The condition became celebrated, once, when the tin organ pipes in a St. Petersburg cathedral collapsed in the depth of a Russian winter.

Scientists were sure that another startling effect of low temperatures would be the liquefaction of substances that were ordinarily in gaseous form—even when that liquefaction did not occur in nature.

The major gases of the atmosphere are the only ones that occur naturally in quantity, and they do not liquefy no matter how cold atmospheric temperatures become—even in Antarctica.

Chemists, however, could produce gases not found in quantity in nature, and these might liquefy if the temperature was lowered sufficiently, even if that lowering would not suffice for atmospheric gases such as oxygen or nitrogen. Then, too, pressure might help, since that would force the molecules of the gas more closely together and might encourage the process of liquefaction.

The first to go about the process of liquefying gases systematically was the English chemist Michael Faraday (1791–1867).

He used a strong glass tube which he had bent into a boomerang shape. In the closed bottom, he placed a substance that when it was heated would liberate the gas chlorine. He then sealed the open end of the tube.

He placed the end with the chlorine-producing material in hot water, and the other end in crushed ice. Chlorine was produced in greater and greater quantity at the hot end, and so within the closed tube it came under its own increasing pressure. At the cold end, its temperature was lowered, and under the combination of decreasing temperature and increasing pressure, liquid chlorine was produced, at that cold end, in 1823.

This demonstrated the principle but established no new horizon. Once liquid chlorine was obtained, its boiling point could be measured, and it turned out to be 238.5° K. (−34.5° C.). If one had a sealed container of chlorine gas, a very cold winter day in Moscow or Montana would suffice to liquefy it.

Liquefied gases can be used to lower temperatures further. Suppose a gas has been liquefied under pressure. If the pressure is then slowly decreased, the liquefied gas begins to vaporize. This means that the molecules of the liquid must move apart to form the vapor, and to do that they must gain energy. If the liquid is kept in an insulated chamber, little of that energy can be gained from the outside world. It must, instead, be gained from the only substance accessible—the liquid itself. As the liquid evaporates, then, the temperature of that portion that remains unevaporated drops.

In 1835, a French chemist, C.S.A. Thilorier, used the Faraday method to form liquid carbon dioxide under pressure, using metal cylinders which would bear greater pressures than glass tubes would. He prepared liquid carbon dioxide in considerable quantity and then allowed it to escape from the tube through a narrow nozzle. The rapid release of pressure evaporated some of the emerging liquid and lowered the temperature of the rest so sharply that the carbon dioxide froze.

Liquid carbon dioxide is stable only under pressure. Solid carbon dioxide exposed to ordinary pressures will sublime, evaporating into the gaseous form directly, without melting. (That is why solid carbon dioxide is called "dry ice.") The sublimation point is 194.5° K. This means that on the very coldest nights in the coldest part of Antarctica, a quantity of carbon dioxide gas would freeze, but only just barely.

Thilorier went further. He mixed solid carbon dioxide with diethyl ether (the common anesthetic), which remains liquid even at the temperatures of solid carbon dioxide. Even at that low temperature, however, the diethyl ether tends to evaporate, though far more slowly than it does at ordinary temperatures. As it evaporated, the temperature of the mixture dropped further. Thilorier managed to reach temperatures as low as 163° K.—and even Antarctica was surpassed. For the first time, the lowest temperatures on Earth were of human origin.

Making use of this particularly cold mixture, other gases were liquefied—but not all of them!

Faraday found a number of gases he could not liquefy even with this cooling agent plus pressure, and he named them "permanent gases." After Faraday's time, several more gases were discovered which Faraday would have added to the list had he known them and had a chance to study them.

There are a total of eight gases which cannot be liquefied at 163° K., even with pressure. They are oxygen, argon, fluorine, carbon monoxide, nitrogen, neon, hydrogen, and helium. Of these, seven are elements. The one compound is carbon monoxide, the molecule of which consists of an atom of carbon and an atom of oxygen.

In 1869, the Irish chemist Thomas Andrews (1813–85) showed that pressure would liquefy gases only below a certain "critical temperature." Above that critical temperature, pressure would not help. The eight permanent gases were those with critical temperatures below 163° K. In order to liquefy those gases, some way of cooling them down to their critical temperature or lower was needed before pressure could usefully be applied.

Again scientists turned to cooling through expansion. This principle did not apply only to the evaporation of a cold liquid. A gas if allowed to expand under conditions where heat could not leak in from the outside world would also cool. This was known as the "Joule-Thomson effect" because it had been worked out by Thomson (Lord Kelvin) in collaboration with his friend the English physicist James Prescott Joule (1818–89).

The first to attempt to use the Joule-Thomson effect to achieve new lows in temperature was the French physicist Louis Paul Cailletet (1832–1913) in 1877.

He began by strongly compressing oxygen in an insulated container. This meant that the gas grew hot in a reverse Joule-Thomson effect. He then bathed the tube containing the compressed oxygen in cold water to extract and carry off as much heat as possible. He next allowed the cold oxygen to expand very rapidly. Its temperature dropped precipitously, and Cailletet finally managed to cool it to the point where he obtained a fog of liquid droplets on the walls of the container. The droplets were liquid oxygen. Naturally, the fog disappeared quickly as heat leaked slowly into the tube, but it had been there.

He managed to do the same, later on, for nitrogen and carbon monoxide, which required even lower temperature for liquefaction than oxygen did.

Working at the same time as Cailletet, but independently, the Swiss chemist Raoul Pierre Pictet (1842–1929) also turned the trick.

He used a slightly different method, cooling compressed oxygen with liquid carbon dioxide. He managed to get the oxygen at a pressure of several hundred atmospheres and a temperature of 133° K., which was well below its critical temperature. When he opened an escape valve to the tube containing the oxygen, the compressed gas escaped, expanding so rapidly that the temperature dropped to the point where a jet of liquid oxygen spurted out and, of course, rapidly evaporated.

In 1883, two Polish chemists, Karol S. Olszewski (1846–1915) and Zygmunt F. Wroblewski (1845–88), improved on the methods of the earlier workers and were able to produce liquid oxygen in quantity. They surrounded the tubes containing the liquid oxygen with other liquids that were as cold as possible in order to cut down the rate of evaporation. As a result, they could for the first time accumulate the liquid forms of the no-longer-permanent gases and study them at leisure.

Olszewski (after Wroblewski's death in a laboratory accident) went on to prepare, in quantity, liquid nitrogen and liquid carbon monoxide. Argon and fluorine had not yet been isolated, but when they were, the same methods could be used to liquefy them.

It turned out that oxygen has a boiling point of 90.2° K., argon 87.5° K., fluorine 85.1° K., carbon monoxide 81.6° K., and nitrogen 77.3° K.

With these gases liquefied, scientists had reached temperatures characteristic of the visible surfaces of such outer planets as Uranus. It was discovered in 1981 that Titan, the largest satellite of Saturn, has a dense atmosphere that consists largely of nitrogen, and there is a general feeling that it may have lakes and rivers on its surface—of liquid nitrogen.

(At the temperature of liquid nitrogen, in fact, one of the so-called permanent gases could actually be frozen. The freezing point of liquid argon is 83.9° K.)

By 1895, then, the only gases that remained unliquefied were neon, hydrogen, and helium, and of these, neon and helium had not yet been discovered (though they would be, in the course of the next three years).

The next great attack, then, was on hydrogen.

As in the case of pressure, the Joule-Thomson effect does not work in liquefying gases until the particular gas being worked on has already reached a particular stage of coolness. For all gases but neon, hydrogen, and helium, it was not very difficult to reach the point where the Joule-Thomson effect worked. In fact, room temperature was cool enough, and to spare, for all other gases.

For hydrogen, however, this was not so. The Joule-Thomson effect did not work for hydrogen till it had been cooled down to 190° K. This had to be understood before hydrogen could be liquefied, and the first to do so was a Scottish chemist, James Dewar (1842–1923).

To begin with, Dewar cooled his hydrogen by plunging his container of that gas into a bath of liquid nitrogen. With the hydrogen now well below 190° K., but still gaseous, Dewar began the process of contraction and expansion, and in this way finally produced liquid hydrogen. It turned out that liquid hydrogen has a boiling point of 20.3° K.

The production of liquid hydrogen meant that scientists on Earth had finally succeeded in working at a temperature lower than any existing on any world of the solar system. The surface temperature even on distant Pluto at its most distant is high enough, thanks to the feeble radiation from a sun 7,300,000,000 kilometers away, to keep any neon and hydrogen (if present) gaseous.

To be sure, Jupiter and perhaps the other giant planets seem to be made up of hydrogen dense enough to be liquid in its properties, but that is liquid hydrogen under enormous pressures and at white-hot temperatures. There is no liquid hydrogen anywhere in the known solar system that is liquid through low temperature alone and under little or no pressure—except in earthly containers.

At liquid hydrogen temperatures, all other substances (but one) are themselves solid. Of the gases that had been liquefied in the last quarter of the nineteenth century, oxygen has a freezing point of 54.7° K., argon 83.9° K., fluorine 53.5° K., carbon monoxide 74° K., nitrogen 63.3° K., and neon 24.5° K.

The only liquid substance at 20° K. is hydrogen itself, and the only gas is helium.

If liquid hydrogen is allowed to evaporate under conditions where heat cannot reach it from outside, the unevaporated portion of the hydrogen will be cooled well below its liquefaction point and will finally freeze at 14.0° K.

And *still*, even at that low temperature, helium remains stubbornly gaseous. It is the one gas in a temperature world of universal solidity.

Dewar failed to liquefy helium, and the problem was taken over by a Dutch physicist, Heike Kamerlingh-Onnes (1853–1926). He began by cooling a sample of gaseous helium in a bath of liquid hydrogen. Only then did the temperature of the helium gas drop low enough for the Joule-Thomson effect to take over. In 1908, Kamerlingh-Onnes liquefied helium at a temperature of 4.2° K.

## TOWARD ABSOLUTE ZERO

In a way, even this was not the end of gases, though in 1908 it certainly appeared as though it was. As the twentieth century wore on, it was recognized that atomic varieties, or isotopes, existed.

Thus, hydrogen consists, in nature, of two stable isotopes, "hydrogen-1" and "hydrogen-2." Atoms of hydrogen-1 possess nuclei consisting of a single proton; atoms of hydrogen-2 possess nuclei consisting of a proton plus a neutron.

Protons and neutrons have about equal masses, so that hydrogen-2 is twice as massive as hydrogen-1, and hydrogen-2 atoms are therefore more difficult to break loose from their fellow atoms and cause to evaporate. In general, this is true for all elements. The isotopes with more particles in the nuclei have higher boiling points than do those with fewer particles.

In most elements, the number of particles in the nuclei is so great, in any case, that one or two particles, more or less, makes little difference. In the case of hydrogen, the difference between a nucleus containing one particle and another containing two is 100 percent. While ordinary hydrogen (in which 6,999 atoms out of every 7,000 are hydrogen-1) has a boiling point of 20.3° K., the boiling point of hydrogen-2, if a sizable quantity is isolated and tested, is 23.4° K., more than 3 degrees higher than that of hydrogen-1.

The case of helium is reversed. Almost every helium atom has a nucleus made up of two protons and two neutrons and is, therefore, "helium-4." It is that which has a liquefaction point of 4.2° K. One out of every 750,000 helium atoms, however, has a nucleus made up of two protons and one neutron, and is "helium-3."

If a quantity of helium-3 were isolated in pure form, and if a container of it were bathed in liquid helium at 4.2° K., the helium-3 being the lighter of the two, would still remain gaseous. Helium-3 would have to be cooled down an additional degree, for it would not liquefy until a temperature of 3.2° K. had been reached.

Kamerlingh-Onnes did not suspect the existence of helium-3, but that did not matter. By allowing helium to evaporate under insulating conditions, he reduced its temperature to 0.83° K. before he died. This meant he had reached a temperature at which no substance could be entirely gaseous, though even at *that* temperature, some helium would exist as a vapor in equilibrium with the liquid.

Once temperatures of less than a degree above absolute zero had been produced in the laboratory, it turned out that humanity had accomplished something remarkable.

I said earlier that no known planet in the solar system was cold enough to liquefy hydrogen under conditions of little or no pressure. There might, however, be planets farther from the sun than Pluto is. There are speculations, for instance, that vast numbers of comets exist far beyond Pluto.

These very distant bodies ought to be colder than Pluto. Might they be so cold that hydrogen would freeze upon them, so cold that even helium would liquefy and freeze? Suppose an object were lost deep in the space between the galaxies, so far from any star that it received virtually no starlight at all. How cold would it then be? Would it be at absolute zero?

To suppose that such bodies would be at absolute zero would seem a reasonable conclusion, but it would be wrong. The universe is full of cosmic radiation as well as of a background of microwave radiation that reaches every corner of its volume. Any object in the universe, no matter how nonproductive of energy it is, and no matter how far it is from any star or any other obvious producer of energy, still absorbs enough cosmic rays and microwaves to reach a temperature of 3° K. That is the general temperature of the universe.

This means that (provided there are no other intelligent organisms in far-distant places who are also experimenting with extremely low temperatures) Kamerlingh-Onnes produced temperatures with liquid helium that were a *universal* record. He had produced temperatures lower than any that existed in nature *anywhere in the universe*.

And yet in one respect, he had failed. He never achieved the final victory. Even at a temperature of 0.83° K., helium remained a liquid. Kamerlingh-Onnes could not achieve the goal of producing solid helium.

Nor could he have, however low a temperature he attained. As it happens, at absolute zero the energy content of atoms and molecules is not quite zero. If it were, and if helium atoms (or any other kind) had zero energy, then they would be motionless. Their momentum would be zero and their position could be exactly determined.

There exists, however, a very powerful "uncertainty principle," first worked out by the German physicist Werner Karl Heisenberg (1901–76) in 1927, to the effect that position and momentum can never be simultaneously and perfectly determined, but that there must always be a residual uncertainty. Some very tiny quantity of energy is retained by matter even at absolute zero, and can never be removed, in order to maintain the requirement of the uncertainty principle.

Since this tiny bit of energy can never possibly be removed, absolute zero is indeed absolute. Since that last bit of energy is there, however, and is sufficient to keep helium liquid, helium remains liquid right down to absolute zero.

At least, this is true at ordinary atmospheric pressure. In 1926, a few months after Kamerlingh-Onnes died, his co-worker Willem Hendrik Keesom (1876-1956) applied pressure to liquid helium and solidified it at last.

It was the final goal. Conditions had finally been reached in the laboratory (and perhaps nowhere else in the universe) in which every substance was in the solid state.

With that, the game seemed to be over, for what else was there to accomplish with temperatures extending down to a fraction of a degree above the absolute zero and with everything solid?

Actually, as it turned out, some types of matter revealed very unusual properties in the neighborhood of absolute zero. Mercury, lead, and a variety of other metals and alloys, at certain crucial and very low temperatures, lost *all* resistance to an electric current and became "superconductive." There were other peculiar properties possessed by liquid helium, particularly.

In order to study all these properties, scientists continued to push for lower and lower temperatures. Nor were they under any illusion that, having attained temperatures less than 1° K., there remained only a quick last push and the ultimate victory. In 1906, the German chemist Walther Hermann Nernst (1864-1941) had worked out what is called the "third law of thermodynamics." From that third law, one can deduce that halving the absolute temperature always takes the same effort regardless of the starting point. No amount of effort, therefore, will reach absolute zero, any more than it will reach the velocity of light.

Nevertheless, scientists saw value in approaching absolute zero as closely as they could.

The technique of allowing gases to expand and liquids to evaporate had reached a dead end in the 1920s at about 0.5° K. Something new was needed.

In 1926, the Dutch chemist Peter J. W. Debye (1884-1966) and the American chemist William Francis Giauque (1895-    ) independently suggested a new technique.

In certain substances, such as gadolinium sulfate, the metal atoms, gadolinium in this case, act like tiny magnets. In the presence of a strong magnetic field, all the atoms line up in one direction and the substance is magnetized. If the magnetic field is then removed, the atoms jiggle around randomly and the salt loses its magnetic properties.

The atoms, in beginning to move randomly, absorb heat and, if insulated from the outside world, must absorb it from the substance itself, so that its temperature falls. If, then, gadolinium sulfate is magnetized and then cooled to the lowest possible temperature, and the magnetic field is then removed, the temperature drops still lower.

It took a while to make the technique work, but by 1933, Giauque used gadolinium sulfate to produce a temperature of 0.25° K. In the same year, Dutch chemists used cerium fluoride to obtain a temperature of 0.13° K. and cerium ethyl sulfate to obtain one of 0.0185° K. Since then, the use of the technique has produced temperatures as low as 0.003° K.

In 1962, the German-English physicist Heinz London (1907–      ) pointed out that helium-3 atoms, being lighter than helium-4 atoms, move more quickly and contribute disproportionately to the temperature. At very low temperatures, helium-3 and helium-4 do not mix perfectly and techniques can be used to abstract the former from the mixture, taking with it most of the heat, and thus lowering the temperature of what remains still further.

A combination of the helium-3 and demagnetization methods has thus produced temperatures as low as 0.00002° K.—temperatures within a fifty-thousandth of a degree of absolute zero.

# 23
# Luminosity

## MAGNITUDE

Changes in temperature produce a number of effects, and in this chapter, we will consider one—a very noticeable one.

All substances, at any temperature above absolute zero, give off waves of electromagnetic radiation. The pattern of wavelengths is characteristic of the particular temperature, and so the temperature of a distant star can be determined from the pattern of wavelengths which it gives off.

In general, the wavelengths cover a large spread from quite long to quite short, with a peak at some intermediate value. As the temperature goes up, the wavelengths shift toward the shorter (and, therefore, more energetic) end of the spectrum, and so does the peak.

For all ordinary temperatures, up to about 600° C., the radiation given off is in the radio-wave region, or the somewhat shorter microwave region, or the still shorter infrared region. All these groups of radiation are marked by one thing in common—they do not affect the retina of the eye. Thus, a block of steel at any temperature in the 1,000-degree range between absolute zero and 600° C. may feel cold, warm, or even too hot to touch, and may give off floods of radiation; but none of that radiation will be sensed by our eyes, and in the dark that block of steel will remain invisible.

Above 600° C., however, some of the radiation spreads out into the visible-light region. At first, only the longest waves of visible light radiate in perceptible quantity, and these, being red in color, make the object "red-hot." As temperature rises farther, more and more of the shorter wavelengths of light appear, and the object becomes orange and then whitens and becomes "white-hot," as all the colors of the spectrum are included. An object as hot as the sun's surface radiates chiefly in the visible light spectrum, though the sun's radiation spreads out well beyond that into the invisible longer waves and the invisible shorter waves.

We associate high temperature with a visible glow, and the only way of keeping a very hot object invisible is to enclose it with something colder. Thus, the very hot interior of Earth is invisible to us only because of the enclosure of the cold crust.

It is also possible to have a source of light that is not at high temperature. Fireflies are a well-known example of that.

To human beings generally, the great source of light is the sun and, by extension, the moon and the other astronomical objects as well. To be sure, even in early times, there was fire. Human beings learned how to maintain and make fires of their own, and eventually artificial illumination became more important to humanity than the light of any astronomical object other than the sun itself. Nevertheless, the comparative brightness of the astronomical objects has always been of great interest to scientists and still remains so.

In early times, there were only two types of properties one could work out for the stars without modern instruments, and that only roughly. These were position and brightness.

The first person we know of who tried to map the heavens and indicate the position of at least some of the various stars was the Greek astronomer Hipparchus (190–120 B.C.). About 130 B.C., he prepared a map on which he listed 1,080 stars, giving the celestial latitude and longitude of each, as best they could be determined without a clock, telescope, or any other modern instrument.

Nor did he neglect the other property—brightness. Hipparchus divided the stars into six classes. The first class included the twenty brightest stars in the sky. The second included stars a bit dimmer than those, the third stars still dimmer, and so on. The sixth included those stars just barely visible on a dark, moonless night to a person with sharp vision.

Each class eventually came to be called a "magnitude" from the Latin word for "large," since it was assumed that all the stars were at equal distances, fixed to a solid "firmament," and that the ones that shone more brightly did so because they were larger. The brightest stars, therefore, were of the "first magnitude," the next brightest were of the "second magnitude," and so on. We retain that system to this day.

The division of the stars into magnitudes was purely qualitative at first. Some first-magnitude stars are clearly brighter than other first-magnitude stars, but no account was taken of that. Nor did early astronomers worry overmuch that the dimmest first-magnitude stars were not very much brighter than the brightest second-magnitude stars. There is, in fact, a continuous gradation of brightness among the stars, but the classification into discrete classes obscures that.

In the 1830s, attempts began to be made to improve on the 2,000-year-old Hipparchian system at last.

One pioneer was the English astronomer John Herschel (1792–1871), who was observing the southern stars from southernmost Africa. In 1836, he devised an instrument that would produce a small image of the full moon that could be brightened or dimmed by the manipulation of a lens. The image could then be made equal in brightness to the image of a particular star. In this way, Herschel could estimate the comparative brightness of stars quite finely, and could determine gradations smaller than a whole magnitude.

Using the full moon, however, restricted the times when measurements could be made and allowed only the brighter stars to be measured, since the dimmer ones were washed out in the moonlight.

At about the same time, however, a German physicist, Karl August von Steinheil (1801–70), had worked out a similar device that could bring into juxtaposition the images of two different stars, one of which could be dimmed or brightened to match the other. Now, at last, magnitudes could be made quantitative.

Steinheil felt that gradations in brightness were sensed logarithmically. That is, the eye was affected by ratios of brightness rather than by actual differences. Thus if star A was three times as bright as star B and nine times as bright as star C, star B would look as much brighter than star C as star A looked brighter than star B. In each case, the ratio was three.

In 1856, the English astronomer Norman Robert Pogson (1829–91) pointed out that the average first-magnitude star is about a hundred times as bright as the average sixth-magnitude star, according to the careful measurements that were now possible. In order to make the five intervals between the six magnitudes come out to just 100, we must make the ratio of each of the five intervals the fifth root of 100, which comes out to about 2.512. (In other words, 2.512 × 2.512 × 2.512 × 2.512 × 2.512 is just about equal to 100.)

Therefore, if you choose a magnitude of 1.0 in such a way as to place it more or less midway among the first-magnitude stars, you can then proceed to work your way down by ratios of 2.512. As photometers improved, astronomers could determine magnitudes to one decimal and even, on occasion, could make a stab at the second decimal. The brighter of two stars, separated by a tenth of a magnitude, is about 1.1 times as bright as the dimmer one; or, if separated by a hundredth of a magnitude, 1.01 times as bright.

Using the new system, we no longer have to limit ourselves to saying that Pollux and Fomalhaut are both first-magnitude stars. Instead, we can say that Pollux has a magnitude of 1.16 and Fomalhaut one of 1.19.

By the time Pogson had worked out his magnitude scale, the sixth-magnitude stars were by no means the dimmest that could be seen. The telescope revealed far dimmer stars, and successive improvements of the instrument revealed still dimmer ones. By continuing the ratio of 2.512, we can have seventh-magnitude stars, eighth-magnitude stars, ninth-magnitude stars, and so on, measuring each to as close a value as our instruments will allow us to.

The best contemporary telescopes will reveal stars as dim as the twentieth magnitude if we place our eye to the eyepiece. If we place a photographic plate there instead, and let the focused light accumulate, we can detect stars down to the twenty-fourth magnitude.

A twenty-fourth-magnitude star is eighteen magnitudes dimmer than the dimmest object we can see with the unaided eye. By the logarithmic scale, this means that the dimmest star the ancients could see is about 16,000,000 times as bright as the dimmest star *we* can see.

There are, of course, stars that are brighter than the average first-magnitude stars and that, therefore, have magnitudes less than 1.0 (the lower the number, remember, the brighter the star). Thus, we have Procyon, with a magnitude of 0.38.

Procyon is not, however, the brightest star in the sky. There are a very few stars that are actually brighter in magnitude than 0.00, and that have negative magnitudes. There is Alpha Centauri with a magnitude of −0.27; Canopus with one of −0.72; and Sirius with a magnitude that goes past the −1 mark, for its magnitude is −1.42.

This means that Sirius, the brightest of the traditional first-magnitude stars, is actually three magnitudes brighter than the dimmest traditional first-magnitude star, which is Castor, with a magnitude of 1.58. Sirius is about sixteen times as bright as Castor and is about 15,000,000,000 times as bright as the faintest star our telescopes can show us.

Are there objects in the sky that are brighter than Sirius?

Certainly! Several of the planets are brighter than Sirius at least some of the time. When the planet Jupiter is at its brightest in the sky, it reaches a magnitude of −2.5. Mars can reach a magnitude of

—2.8, while Venus, the brightest of all the planets, can attain a magnitude of —4.3. At its brightest, Venus is about fifteen times as bright as Sirius.

Even that doesn't represent the top. The moon is far brighter even than Venus, and when the moon is full, it can attain a magnitude of —12.6. That means the full moon is about 2,000 times as bright as Venus.

That leaves the sun, whose magnitude is —26.91. The sun is 525,000 times as bright as the full moon, 1,000,000,000 times as bright as Venus, 15,000,000,000 times as bright as Sirius, and fully 25,000,000,000,000,000,000 times as bright as the dimmest object the telescope will show us.

Since there is nothing brighter to see in the sky than the sun and nothing dimmer than the dimmest star telescopes can show us (at least until telescopes are further improved), we have reached the limit in both directions, having traversed a range of fifty-one magnitudes.

## ABSOLUTE MAGNITUDE

So far, all the magnitudes being discussed are *apparent* magnitudes. The brightness of an object, as seen by us, depends not only on how much light it emits, but also on how distant it is from us. An object that is actually extraordinarily dim in an absolute sense, such as a 100-watt light bulb, can be placed just in front of us and seem brighter to our eyes than the moon does. On the other hand, a star that gives out much more light than the sun can be so far away that not even a telescope will show it to us.

In order, then, to determine levels of *real* brightness, to measure the light an object *actually* emits—its "luminosity"—we must imagine that all the objects in question are at some fixed distance from us. The fixed distance has been selected (arbitrarily) as 10 parsecs, a distance equal to 32.6 light-years.

Once the distance of any luminous object is known and its brightness at that distance is measured, we can calculate what its brightness would be at any other distance. The magnitude an object would have if it were exactly 10 parsecs distant from us is its "absolute magnitude."

Our sun, for instance, is about 149,500,000 kilometers from us, or 1/200,000 of a parsec. Imagine it 10 parsecs from us, and you have increased its distance by 2,000,000 times. Its apparent brightness sinks by the square of that number or 4,000,000,000,000 times. That means its brightness sinks by about 31.5 magnitudes. Its absolute magnitude is thus about 4.7.

The sun, seen from a distance of 10 parsecs, would be visible, but it would shine as a fairly dim and quite unremarkable star.

What about Sirius? It is already at a distance of 2.65 parsecs. If we imagined it out at 10 parsecs, its brightness would dim by nearly three magnitudes, and its absolute magnitude would be 1.3. It would no longer be the brightest star in the sky, but it would still be a first-magnitude star.

The absolute magnitudes, which wipe out difference in distance as a factor, show us that Sirius is about twenty-three times as luminous as the Sun; that is, it emits twenty-three times as much light.

Sirius is far from the most luminous star there is, however. There are stars that do much better. Of all the first-magnitude stars, the most distant is Rigel, which is 165 parsecs away. It is only the seventh-brightest star in the sky and is only one-quarter as bright as Sirius. Still, Rigel is over sixty times as far from us as Sirius is. To make so respectable a show from such a distance, Rigel must be very luminous.

And indeed it is! The absolute magnitude of Rigel is −6.2. Place it at a distance of 10 parsecs, and even though it would be at nearly four times the actual distance of Sirius, it would not only greatly outshine that star, it would shine even brighter than Venus—about six times as bright. In fact, Rigel is 1,000 times as luminous as Sirius and 23,000 times as luminous as the sun.

Rigel is the most luminous star that we know of in our own galaxy, but there are other galaxies. The Larger Magellanic Cloud is a satellite-galaxy of our own and in it is a star called S Doradus. It is too dim to see except with a telescope, but it is some 45,000 parsecs away and astronomers were astonished it was as bright as it appeared, considering its distance. It turns out to have an absolute magnitude of −9.5. That makes it about twenty-one times as luminous as Rigel and nearly 500,000 times as luminous as the sun.

If S Doradus were in place of our sun, a planet circling it at seventeen times the distance of Pluto would see it shine as brightly as we see our sun shine.

S Doradus is the most luminous stable star we know of. It emits more light, day after day, century after century, than any other. Not all stars are stable, however. Occasionally, stars explode and, in the process, gain luminosity sharply, if temporarily.

The size of the gain depends on the size of the star. The more massive the star, the more enormous the explosion. The really magnificent explosion of a "supernova" can bring a single massive star to an absolute magnitude, very briefly, of about −19.

For a brief time, such a supernova will be shining with a luminosity some 6,000 times that of S Doradus, and about 10,000,000,000 times that of our sun. Even at a distance of 10 parsecs, it will shine 360 times as brightly as the full moon, and be a thousandth as bright as our very nearby sun.

Have we now reached the upper horizon of luminosity? No!

A supernova is only a single star. Might we not consider the luminosity of a group of stars?

A pair of stars, reasonably close together, looks like a single star from a distance. If both stars are of equal brightness, the combination is 0.75 magnitudes brighter than either star singly.

Double stars are very common, and even triple or quadruple star systems are not exactly rare. In fact, stars exist in large clusters as well. There are about 125 known "globular clusters" associated with our galaxy, and each contains anywhere from ten thousand to several hundred thousand stars, all densely packed together (at least densely packed by the standards of our own stellar neighborhood).

Suppose, then, we consider a globular cluster made up of 1,000,-000 stars, each with the luminosity of our sun. Such an enormous cluster would be, nevertheless, merely twice as luminous as the single star S Doradus. A gigantic supernova can attain a luminosity equal to 3,000 times that of a large globular cluster. No globular cluster can, therefore, set a luminosity record.

A galaxy itself, however, has a nucleus which is the equivalent of a globular cluster of enormous size. The center of our own galaxy is a densely packed cluster made up of stars which, in total, emit 100,-000,000,000 times the light of our sun. Its absolute magnitude can be calculated to be −22.8. (The rest of the galaxy, outside the nucleus, has a relatively sparse scattering of stars, and if its luminosity is included, the overall value for the absolute magnitude may reach −22.9.)

*That* looks like a new record. The galactic nucleus shines with a luminosity over three times that of a supernova at its peak. There are galaxies larger than our own, too. A large galaxy can easily be ten times the mass of our own and have an absolute magnitude of −25.

There is a catch to this calculation of the absolute magnitudes of globular clusters and galaxies, however, for we are dealing with extended bodies. A large globular cluster could be up to 100 parsecs across, and a galactic nucleus can be up to 5,000 parsecs across. The absolute magnitude can be calculated, but it can't be experienced in the ordinary way.

If you imagined the central point of a globular cluster or of a galactic nucleus to be 10 parsecs away, you would be *within* the object. You would see stars all about you and you would not have the sense of a combined luminosity, any more than you have it now in our own galaxy.

To be sure, we might use 1,000,000 parsecs as the conventional distance in measuring luminosity, and then we would see that a large galaxy would outshine any individual star under any circumstance. But then, all objects seen at that distance (even in a good telescope) would seem very dim and unimpressive.

If we want to look for a luminosity record beyond a supernova, we must ask if there is anything that would look like a single object of reasonably small size at a distance of 10 parsecs and that would yet outshine a supernova from day to day, steadily.

There *is* such a thing. What we call "quasars" are, apparently, galactic nuclei so condensed and so brilliant that they can be seen (telescopically) at distances of hundreds of millions of parsecs. No other object can be seen at such distances. A typical quasar is thought to be, perhaps, only half a parsec or so in diameter, and yet it shines with the luminosity of a hundred galaxies such as our own.

Half a parsec is a respectable diameter; it is about 12,000,000 times the diameter of our sun; it is over 1,000 times the diameter of the orbit of Pluto. Place a quasar at a distance of 10 parsecs and its apparent diameter would be nearly 3 degrees across. That is about six times the apparent diameter of our sun or full moon, but we would still see it as a single blazing object—rather as we would see our own sun from the surface of the planet Mercury. (There would also be individual stars outside that brilliant nucleus, but they would play no significant part in the matter of luminosity.)

The average quasar will then have an absolute magnitude of −28. It will shine, *even at the distance of 10 parsecs,* about twice as brilliantly as the sun does in our sky, even though the quasar is 2,000,000 times as far away.

But how bright is the *brightest* quasar?

In 1975, two Harvard astronomers studied past photographs of quasar 3C279. It usually shines with an apparent magnitude of 18, but back in 1937 (when no one knew that the apparent dim and undistinguished star was actually an enormous and extremely distant quasar) it briefly attained an apparent magnitude of 11.

To shine as brightly as the eleventh magnitude from a distance of about 2,000,000,000 parsecs is almost incredible. At its peak, 3C279 shone with the light of 10,000 ordinary galaxies and its absolute magnitude was calculated to have reached a peak of −31.

Imagine 3C279 at a distance of 10 parsecs from us and it would shine with a brilliance forty times that of our sun as we see it now. A quasar such as 3C279 can reach a peak luminosity, then, of 100,000,-000,000,000 times that of our sun—or 500,000,000 times that of S Doradus—or 60,000 times that of a vast supernova at its peak.

And that is the most luminous object we know of, as yet.

In this book I have followed a few of the remarkable expansions of horizons that human daring and modern science have brought about. There are other horizons that have expanded—magnetic intensity, viscosity, angular momentum, etc.—but those I have presented are sufficient to show humanity at its most magnificently human in its restless desire to push on as far as possible in all directions.

# Index